THE SCAM SELF-DEFENCE HANDBOOK

By Kevin O'Flaherty

THE SCAM SELF-DEFENCE HANDBOOK

Australia Edition 2026

Copyright © 2026 Kevin O'Flaherty

All rights reserved.

First published in 2026 by

The Everyday Technologist

Tasmania, Australia

Printed and distributed by Ingram Content Group

Cover and interior design by Olivier Darbonville

Disclaimer: This book provides general information only and does not constitute professional advice. The author and publisher disclaim liability for actions taken based on this content.

ISBN-13 (Print): 978-1-7644915-3-2

ISBN-13 (eBook): 978-1-7644915-9-4

Dedicated to my parents, Eleanor & Pierce (Ma and Da)

In fact, this is for all the parents out there!

CONTENTS

SETTING THE SCENE FOR BATTLE: WHY DIGITAL SELF-DEFENCE MATTERS

FEAR NOT!

To some, technology feels like a towering beast – unpredictable, overwhelming, and just a little bit terrifying. It beeps, it buzzes, it throws cryptic error messages and suspicious emails your way. The jargon is nauseating, and the news is terrifying.

So, when scammers enter the scene, it can feel like the whole digital world is out to get you.

If you've ever felt that *Fight*, *Flight*, or *Freeze* response kick in when dealing with technology, you're not alone. That reaction is deeply human. When something feels bigger than us, we run. When it feels *much* bigger, we freeze. We panic. We shake. We shut down.

TECHNOLOGY IS NOT AS BIG AND SCARY AS IT SEEMS.

But here's the truth – one that rarely gets spoken aloud in the tech world:

Technology is not as big and scary as it seems. In fact, you're more than a match for it.

This handbook is your guide to reclaiming your confidence. It's not about becoming a tech wizard overnight. It's about understanding just enough to feel safe, smart, and in control. Because knowledge is power – and reclaiming power is the key to scam protection.

So yes, there will be moments in these pages when the tricks of scammers might make your skin crawl. Their antics will make you angry or disgusted.

But don't run for the hills. Don't freeze. Instead, lean in. Learn. And remember:

Fear Not. You've got this.

THE BATTLE

There is a battle happening right now. It's happening in your inbox, on your phone, in your everyday life. You didn't enlist or volunteer. You didn't pick the fight. But whether you realize it or not, you're on the frontline.

THE FIGHT ISN'T PHYSICAL. IT'S VIRTUAL — BUT VERY REAL.

The fight isn't physical. It's virtual – but very real. The enemy is of course: Scams. As old as humanity but accelerated by technology.

In this book, we're not just learning about them – we're standing up, fighting back, and taking control.

You don't need to be a digital expert. You don't need complex software, technical know-how, or deep industry knowledge.

What you will need:

- Awareness – understanding how scams work, what to watch for, and how small habits build strong defences.

- Instincts – developing the ability to recognize scam characteristics and build reflexive habits to protect yourself and your community.

- Simple action – taking clear, confident steps so you can navigate your digital world.

These are skills everyone can master. Skills that put control back in your hands. Skills you will learn in this handbook.

HOW BIG IS THE BATTLE?

Scams aren't just a nuisance – they're a global crisis. And the challenge is growing.

In 2024 alone, scammers siphoned over $1.5 trillion (AUD) worldwide [1] -targeting individuals, businesses, and even governments. Fraud is no longer just crime; it's a booming economy – a highly organised, technology-driven industry built to deceive.

Developments in technology are enabling scammers to operate faster, more convincingly, and at greater scale than ever before. What once took weeks to plan can now unfold in seconds. With tools that can mimic voices, forge documents, and manipulate social media, scammers are becoming harder to spot and harder to stop.

At the same time, our ability to respond is struggling to keep up. Many people feel overwhelmed by the pace of change. And that's not just a feeling – it's a real issue.

Lack of access, limited training, or simply not knowing where to start can leave everyday people more vulnerable to online threats.

According to Australian Human Rights Commission, almost 1 in 4 Australians are digitally excluded [2] - meaning they lack the access, skills, or confidence to fully participate online. This gap is even wider for older Australians, people in regional areas, and those with lower incomes.

Scammers know this. They rely on confusion, uncertainty, and gaps in knowledge and confidence. And with billions of dollars at stake, they're investing heavily in making their attacks more effective.

HOW THIS BOOK WORKS

Scam Defence Skills – Plus Confidence, Context, and Control

This isn't just a "Scam Defence 101" guide. It's a practical hand-book for building real-world digital reflexes. Yes, you'll learn the basics – but more importantly, you'll develop the instincts and confidence to respond when things get tricky.

A deeper understanding of how the scam industry operates helps you tap into your natural ability to defend yourself.

You'll start to see patterns, spot manipulation tactics, and make smarter decisions – not just because you memorised a list, but because you understand the game. This book is about equipping you with adaptable skills, not just static knowledge – so you can stay safe, stay sharp, and stay in control.

> **THIS BOOK IS ABOUT EQUIPPING YOU WITH ADAPTABLE SKILLS, NOT JUST STATIC KNOWLEDGE.**

The *Skills Chapters* (2 & 3) will help understanding the Do's and Don'ts of Scam Self-Defence. These skills need to be understood in partnership with the *Knowledge Chapters* (4 & 5). Chapters 1 & 6 cover the role that you and your community play in scam self-defence.

Advice and guidance will be clearly laid out for these Skills and Knowledge as well as Red Flags.

SKILL TIPS

These parts focus on building your confidence and capability. You'll learn how to apply what you know in everyday situations, using simple, practical approaches that grow with your experience.

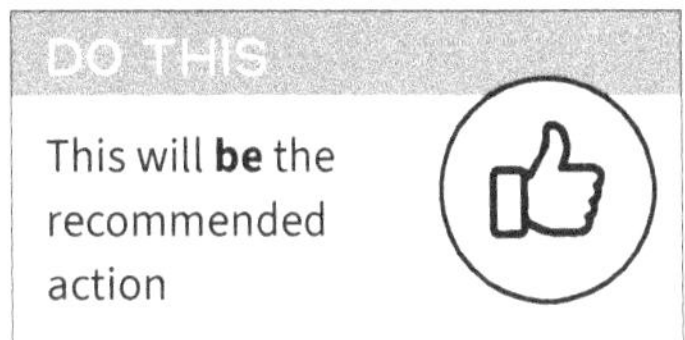

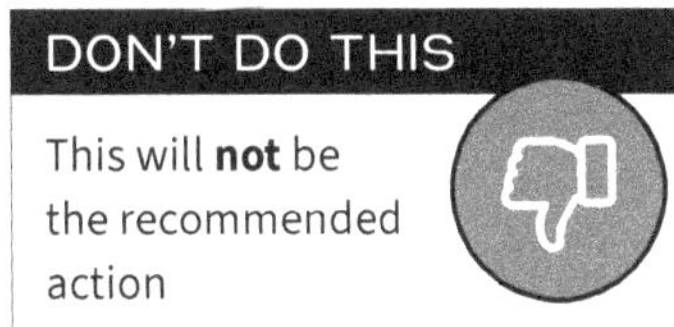

KNOWLEDGE TIPS

These sections explain the ideas and patterns behind scams. How they operate, why they're effective, and what makes them dangerous. Understanding the *why* helps you stay alert and make informed decisions.

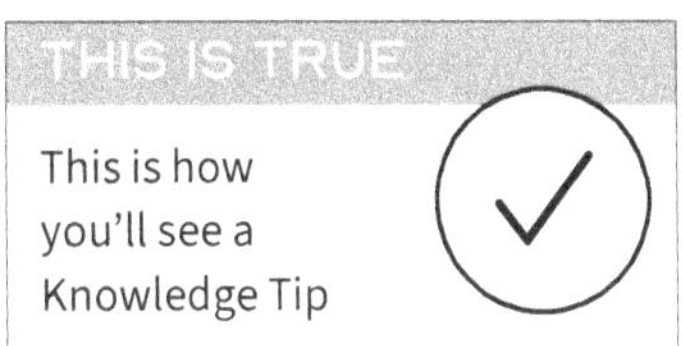

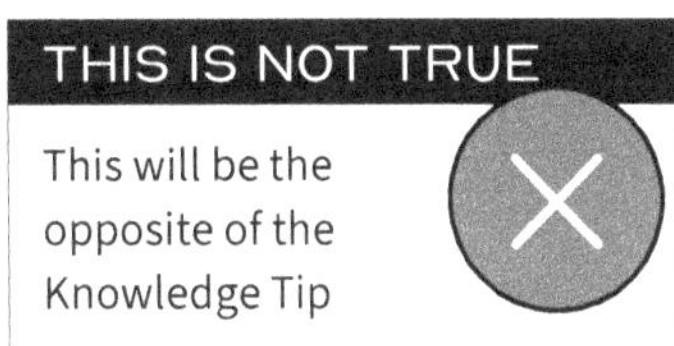

RED FLAGS

These sections highlight the risks using a Red Flag rating. These are the warning signs before they become problems. There is no such thing as 'No Risk' unfortunately, so all scenarios should have a measure of risk awareness. Red Flags will appear like this:

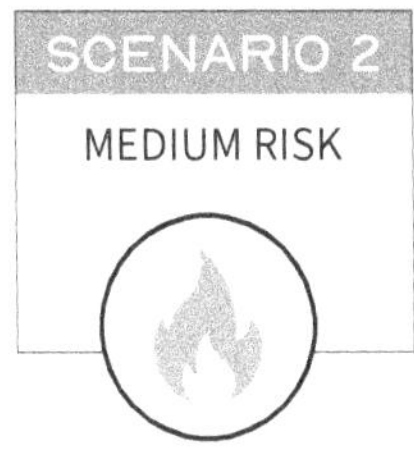

YES/NO AND DO/DON'T

These become tools to help you with simple actions in a cluttered digital world. Giving you clear-thinking choices that work even when you're tired, busy, or unsure.

Each chapter will clearly mark these sections with their respective icons to help you navigate and absorb the content more easily.

There will also be a summary of learnings in table format at the end of each chapter to help you organise your thoughts more clearly as you move through the handbook.

With that, let's start with a couple of key knowledge tips to start with. Afterall, this book is about Scams – so let's define them.

Definition: What is a Scam?
(and what is not a Scam?)

While low-quality products, pushy marketing, and misleading advertising can be frustrating – even unethical in some cases – they are not scams in the way this book defines them.

A scam requires deception with intent to steal money or sensitive information [3]. In contrast, bad business practices may be shady or unfair, but they don't necessarily involve fraudulent intent.

This book focuses on true scams – the kind that actively trick, manipulate, or impersonate others usually for financial harm. Not just frustrating experiences, but outright fraud.

IS A SCAM

- Pretending to be someone they are not.

- Getting you to do something they are not entitled to ask of you.

- Attempting to get money or information from you through deception.

IS NOT A SCAM

- Selling low-quality products or services.

- Providing misleading or pushy marketing.

- Delivering poor standards of customer service.

Definition: The difference between a Scam and a Cyberattack?

I'll add this for clarity as they can often be confused.

A Cyberattack is an attack on computer systems and networks and the information that resides in them, rather than an explicit attack on people.[4]

- Person-to-Person attack
- Uses Human vulnerability
- Requires YOU to be involved

- Person-to-Technology attack
- Uses Technical vulnerabilities
- Doesn't require YOU to be involved.

Note: Not to muddy the waters, but oftentimes a cyberattack starts by scamming. By impersonating someone else, hackers can get your personal information and use that to hack your technology. But for the purposes of keeping-it-simple, we'll maintain a clear separation in the definitions.

Summary

SECTION	KEY CONCEPTS
Scam vs. Cyberattack	Scams target people through deception Cyberattacks target technology through vulnerabilities
Scam Definition	A scam is: An attempt to trick you into handing over your money or personal information A scam isn't: Low quality products, poor customer service or pushy marketing
Navigate the book	Skill Tips are highlighted as Do Don't Knowledge Tips are highlighted as Yes No Risk ratings are Flagged from Low Medium and High
Fight the Fear	Scammers prey on fear. Fight that feeling and read on.

CHAPTER I

SITUATIONAL AWARENESS: MAPPING YOUR DIGITAL BATTLEFIELD

IT'S NOT YOU, IT'S I.T.

Let's be honest: the tech world has made amazing things, but it hasn't always explained them well. Too much jargon. Too much hype. Too many acronyms that leave people feeling confused or shut out. They talk in code when they should be making things clear.

Technology should be built for everyday people and explained in everyday language. That starts with knowing what is (and isn't) a technology problem. Every time you use a device or an app, there are two parts: what the technology does and what you do. Understanding the difference is where control begins.

This matters most when it comes to scams. Here's the truth: scammers don't hack technology, they hack people. The tech part is usually the easy bit. The human part is where the real risk lies…

SCAMMERS DON'T HACK TECHNOLOGY, THEY HACK PEOPLE.

but where your strengths lie too.

Scams are human threats. So they need human defences, starting with strong situational awareness.

Situational awareness is simply knowing how you move through your everyday digital life – the devices you rely on, the habits you've built, the moments where you're most exposed, and the ways scammers try to slide into those ordinary interactions unnoticed.

Before we jump into the practical skills and drills, we're going to take a quick moment to ground ourselves in three simple questions:

EVERYDAY PEOPLE	EVERYDAY TECHNOLOGY	EVERYDAY LIVES

Who are you in the digital world?	What technology are we actually talking about?	Where do these attacks take place in everyday life?

This overview will be brief – just enough to frame the rest of the book. The deeper breakdowns and examples sit in the Appendix if you want to explore them later. For now, let's set the scene so the skills you learn next have something solid to stand on.

EVERYDAY PEOPLE — WHO IS BEING TARGETED IN THIS BATTLE?

This book is for Everyday People – those who use Everyday Technology like phones, email, banking apps, and social media. You're surrounded by a world of digital information, but not all of it feels relevant, safe, or easy to understand. That's normal. In fact, it's the experience of most people on the planet.

But even within this broad group, there are nuances in how people approach technology. Your comfort level, risk tolerance, and willingness to engage-with-change all shape how you experience the digital world and how vulnerable you might be to scams.

To help understand this more and see where you fit in:

Imagine 4 types of people lining up outside a physical store waiting for the next big technology release.

1. The Tech-Enthusiasts

2. The Tech-Willing

3. The Tech-Cautious

4. The Tech-Terrified

Have a look at the descriptions. Where are you in the queue? First in the queue like the Tech-Enthusiasts or maybe not even in the queue at all, like the Tech-Terrified.

Everyday People: Who Are You?

PERSONA	ATTRIBUTES	IS THIS BOOK FOR YOU?
Tech Enthusiasts	Early adopters, confident online, loves new gadgets, high risk appetite	Not for the tech tips -but a reminder on the human side. Maybe you'll buy it as a gift for family :)
Tech Willing	The first of the mainstream adopters. Usually has a positive view on technology.	If you're Tech Willing but not Tech Savvy, this book gives you clearer direction and confidence.
Tech Cautious	Hesitant with tech, prefers guidance, dislikes sudden changes	This book is absolutely for you. It will help you stay cautious - but also aware and empowered.
Tech Terrified	Avoids tech when possible, feels overwhelmed, fears mistakes	This book is absolutely for you. Being online means being confident online and we'll get you there.

EVERYDAY TECHNOLOGY: HOW HAS TECHNOLOGY CHANGED SCAMS?

In 2024, based on Scamwatch's published statistics, around 90% of scams involve some form of digital contact – whether by phone, SMS, email, or online platforms. [3]

Technology has given Everyday People unprecedented access to the world – connecting us, informing us, and streamlining our lives in ways once unimaginable.

But that same access works both ways. While technology opens doors for us, it also creates access back in for scammers to manipulate and deceive. That's why it's important to understand how each communication channel works and why some are more effective for scammers than others.

Scammers have a wide range of digital entry points – from phone calls to social media, email, mobile apps, websites, and SMS. But when it comes to financial impact, some methods prove far more lucrative for fraudsters.

These Technology Channels will be covered in more detail in:

- Chapter 3: Practice Drills
- Appendix: Everyday Technology

But for now, here they are at a glance:

Table: Everyday Technology at a Glance

TECH. CHANNEL	EXAMPLES	WHAT TO WATCH FOR
Phone Calls	Robocalls, fake bank/ government calls	■ Pressure or urgency ■ Requests for action or personal details during the call ■ Unknown or unverified numbers
Social Media	Romance scams, impersonation	■ Strangers claiming to know you ■ Connections asking for money, offering jobs or prizes ■ Threats or bribery claims
Email	Phishing, fake invoices	■ Unexpected payment request ■ Links to click for payment or information ■ Unverified claims, threats or offers
Text or SMS	Delivery or bank alerts	■ Links to suspicious websites ■ Unexpected claims, threats or offers of help ■ Unknown or unexpected sender
Websites	Fake stores, pop-ups	■ Unusual payment options ■ Deals that seem too good to be true ■ Pop-up warnings or error messages asking you to click
Apps	Fake banking or crypto apps	■ Apps asking for excessive permissions ■ No reviews or very poor reviews ■ Apps from unofficial sources (not on App Store or Google Play)

EVERYDAY LIVES: WHERE IS THE BATTLE TAKING PLACE?

Scams can happen anywhere, just as risks exist in the physical world, but some places are safer than others. The important thing is that you already have instincts for this. You have spent your whole life recognising which streets feel safe, which shops you trust, and which parts of town you naturally avoid.

The digital world works in a similar way.

You do not avoid leaving the house simply because certain streets feel risky at night. You still head to the post office midmorning, shop at familiar stores, and move confidently through places you know well. Online life is no different. You do not have to explore every corner of the internet. You just need to stay aware of where you are.

In this book, we will treat your digital world like a town map. Some areas are predictable, some are busy, and some are best approached with caution. Each area has its own rhythms, its own types of people, and its own style of risk. The key is adopting a mindset of Healthy Suspicion. This is not fear or avoidance. It is simply the same everyday awareness you already use in the physical world.

Think of the next page as your map. It is a quick guide to the different neighbourhoods you might travel through online. It is not a rulebook. Instead, it is a way to ground your instincts so you can recognise where you are standing and what you should be mindful of.

From here, the skills and drills that follow will make far more sense.

Your Digital Neighbourhood at a Glance

AREA	EXAMPLES & RISK LEVEL	WHAT TO WATCH FOR
Digital Main Street	**Low Risk** Banks, Government, Big Retail	Fake SMS or emails pretending to be official notices
Town Centre	**Medium Risk** Social Media, Marketplaces	Strangers asking for money, links from unknown profiles
Shopping District	**Medium Risk** Smaller businesses, niche stores	Poor reviews, unclear contact details, insecure payments
Nightlife	**Medium-High Risk** Dating apps, gaming platforms	Emotional pressure, too-good-to-be-true offers
Dark Alleys	**Very High Risk** Illegal streaming sites, unsafe pop-ups	Warnings like "This Site is Unsafe," requests to download

When navigating your Digital World you may want to consider principles for yourself such as:

Tech Enthusiast: Comfortable exploring most areas, but always practicing awareness in busier or higherrisk spaces.

Tech Willing: Confident on Digital Main Street, proceed carefully into the Town Centre.

Tech Cautious: Best to stay on Digital Main Street and only visit the Town Centre with intention and extra checking.

Tech Terrified: Keep to Digital Main Street. Only visit the Town Centre with people you already have a relationship with in the physical world.

Chapter Summary

SECTION	KEY CONCEPTS
Everyday People	**Know who you are and what your risks are** ■ Tech Enthusiasts: Not the target audience for this book ■ Tech Willing: Open to technology, but open to influence ■ Tech Cautious: Best to be hesitant out of awareness than fear ■ Tech Terrified: Avoidant and vulnerable
Everyday Lives	**Know where you are comfortable going** Digital Main Street: Banks, Government, Big Retail Town Centre: social media, marketplaces, new services Dark Alleys: Website marked as 'Unsafe', illegal streaming sites
Everyday Technology	**Be vigilant in all technology 'Channels'** ■ The scams differ in tactics but follow familiar patterns. ■ Be wary of the red-flags with each channel ■ Practice the Top Tips for each category

CHAPTER 2

SELF-DEFENCE BASICS: BUILDING REFLEXES TO RESPOND TO DANGER

REFLEX TRAINING

Scams don't just target your technology – they target your attention, your instincts, and your emotions.

But here's the good news: you already have the tools to defend yourself.

Everyday People – regardless of technical capability – have natural instincts that kick in when something feels off. You pause. You hesitate. You get that gut feeling. These are your scam reflexes. You don't need to learn them – you just need to use them.

> **YOUR SCAM REFLEXES. YOU DON'T NEED TO LEARN THEM—YOU JUST NEED TO USE THEM.**

This chapter is about helping you trust those instincts and use them with confidence. Because when scams happen, they're designed to scatter your thinking and rush your decisions. That's why clear-thinking habits matter. They give you something solid to hold onto when everything else feels uncertain.

We'll explore how to:

- Recognize emotional traps like fear and urgency.

- Use simple decision-making rules that work under pressure.

- Understand your relationship with 'trust' and how to give it out.

- What characteristics to focus on when identifying potentials scammers.

- How to prevent yourself from being a 'Soft Target'

You don't need to become suspicious. You just need to become steady. Reflex Training is about building habits that help you stay calm, clear, and in control – every day.

STOP AND VERIFY — THE DEFENSIVE REFLEX

Whenever I discuss online scams with people who identify as Tech-Terrified, I always say:

Open your computer or phone the same way you would open your front door.

By removing technology from the equation, we can rely on human instinct and life experience to guide our decisions.

If someone knocks on our door – someone we don't know, can't identify, or asks for access to our property or cash – we naturally pause, assess, and apply caution before taking action.

When an email or SMS alert dings: that's a digital 'knock on the door'. Before clicking a link or responding to a message, use the same protective mindset you would apply in real life.

Just as we wouldn't blindly trust a stranger at our door, we shouldn't blindly trust an unknown message, email, or phone call. Scammers thrive on instant reactions – the quicker they get you to act, the less time you have to stop and think.

You may not be a technology expert – but you know people. And that's what this is all really about.

So before we get into the useful skills chapters, remember – if you forget everything, just take a breath and imagine what you would do if this interaction was happening at your front door.

STOP = Using the Defensive Reflex to Retake Control

The act of 'stopping' does more than just pause the interaction – it shifts the power dynamic entirely.

👍	**Control**	Scammers want control. By stopping the engagement, you take control.
👍	**Timing**	When scammers have control, they own the timing and can create a fake sense of urgency. When you have control, you 'own' the timing.
👍	**Decisions**	Scammers want to control the decision and actions. Decisions are now yours to make. *(Doing nothing or disconnecting are valid decisions for you to take)*

For anyone who feels out of control when it comes to technology, simply saying STOP is the most powerful first step you can take.

And don't RESTART until you have verified it is a legitimate request.

But how do I even know they're a scammer?

IT'S NOT PRACTICAL — NOR HEALTHY- TO TREAT EVERY INTERACTION AS A SCAM. YOU SHOULDN'T HAVE TO OPEN EVERY MESSAGE OR ANSWER EVERY CALL WITH SUSPICION. BUT BY BUILDING A SIMPLE HABIT OF PAUSING AT THE START OF ANY DIGITAL INTERACTION, YOU GIVE YOURSELF A MOMENT TO TAKE CONTROL. IF YOUR INSTINCTS SAY STOP, LISTEN TO THEM. STOP IMMEDIATELY.

Before we move on to Step 2 of Stop and Verify, let's take a closer look at what triggers that initial STOP, and why it matters.

You stop at Red Flags

We'll explore these 'Red Flags' in more detail shortly. But for now, here's a quick look at the kinds of signals that should prompt you to STOP. These are the warning signs that tell you something might not be right.

RED FLAG	RATING	WHAT TO WATCH FOR
The Urgency Trap	🔥	Compelling you to act now, or quickly.
The Fear Trap	🔥	Attempts to make you fear you have done something wrong or will be in trouble if you don't act.
Requesting Personal Information	🔥	Any request to share personal information.
Requesting Payment	🔥	Any request to make a payment.
Request to Follow a link	🔥	Any request to follow a link in a message.

VERIFY = The Threat Assessment

Stopping gives you control, but verifying gives you certainty.

When you initiate the verification – by calling your bank directly, visiting an official website, or checking with trusted sources – you stay in control. This is called an Outbound Action using Your Information.

Never validate through information or instructions from **INBOUND** messages.

Always use your own trusted, **OUTBOUND** methods to confirm legitimacy.

DO TRUST	DON'T TRUST
■ Your Information	■ Their Information
■ Outbound Action	■ Inbound Influence

Even if a legitimate provider – like your internet company – contacts you about a missed payment, you have every right to say:

"Thanks for letting me know. I'll call your official switchboard using a verified number to follow up."

Feel empowered to do this confidently and calmly. And in your own time. Remember, you're in control now.

Here are some examples of inbound influence versus outbound action and the information types that could be sent.

 **DO TRUST** ■ YOUR INFORMATION ■ OUTBOUND ACTION	 **DON'T TRUST** ■ THEIR INFORMATION ■ INBOUND INFLUENCE
Navigating to a website manually from a trusted source	A link in an unsolicited email or message
Calling a number saved in your phone or found on an official website	A phone number provided in a suspicious email or text
Going directly to the official website to reset your password	An unexpected request to reset a password via a provided link
Searching for the official company contact details yourself	An email claiming to be from a company but using an unfamiliar domain
Manually entering a trusted web address in your browser	A QR code sent by an unknown sender
Visiting the official profile or website of the organisation directly	A social media message urging immediate action with a provided link

THE URGENCY TRAP: RESISTING THE PUSH TO REACT

The first of the 2 instinctive signs to look out for in a potential scam is the pressure to act quickly.

Scammers thrive on speed. They want you to act before you think, trapping victims in a high-pressure decision loop.

Red Flags of the Urgency Trap:

🔥	**Limited Time Offers**	"You've won a prize, but you must claim it within the **next 10 minutes!**"
🔥	**Suspension Threats**	"Your account will be locked in **24 hours**!"
🔥	**Pressure to pay immediately**	"You must send money **now** or risk penalties!"

*I'm using bold and underlining of urgency terms like 'next 10 minutes or 'in 24 hours' for emphasis – scammers won't be so kind.

No matter how much persuasion someone tries to apply, they are not entitled to do treat you this way and you should disconnect from the engagement immediately.

Urgency triggers a sense of panic, which short-circuits your ability to think clearly. When you feel rushed, you're less likely to question the situation, verify the details and spot the red flags. Scammers exploit this by trying to create a sense of chaos.

You cut it off by choosing to STOP.

THE FEAR TRAP: RESISTING FALSE AUTHORITY

If urgency is the push that scammers use to rush you, fear is the pull that drags you into their trap. It's the second instinctive sign to watch for – a deep emotional tug that makes you feel threatened, cornered, or helpless.

Scammers know that fear clouds judgment. They use intimidating language, official-sounding threats, and worst-case scenarios to make you feel like you're in serious trouble. The goal? To make you freeze, panic, and comply.

Red Flags of the Fear Trap:

🔥	**Legal Threats**	"Pay now or face **legal action**!"
🔥	**Financial Threats**	"Your loan is overdue. Pay immediately or we will take steps to **repossess and sell your home.**"
🔥	**Government Threats**	"You owe back taxes. Pay immediately **to avoid prosecution.**"

These messages are designed to make you feel small and powerless. But here's the truth: no one has the right to treat you this way. Fear is a tool of manipulation, not authority.

When you feel that rising panic – tight chest, racing thoughts, shaky hands – that's your signal. You're in the Fear Trap. And the way out is simple:

STOP. Breathe. Disconnect.

You are not powerless. You are not alone. And you are not obligated to respond to fear.

HOW TO RESIST THE URGENCY & FEAR TRAPS

STOP – Your reflexive instinct should halt all engagements where a time or intimidating related pressure is applied.

DISCONNECT – Digital Main Street, like Banks and the Government will never apply pressure in this way.

BLOCK and REPORT.

- Phone calls: block the number and report it using your phone application.

- Emails: Report as SPAM using your email application.

- SMS: Block and Report SPAM using your messaging application.

VERIFY – If you remain curious about the legitimacy of the message:

- Company validation: Contact the supposed company using a phone number or email address you have sourced.

- Web search: Many known scams are published on legitimate websites such as Scamwatch and Banking Websites.

- Ask someone else: Talk to a friend, family member or neighbour and use them as a sounding-board.

Just like in the real world, 'where you are' matters.

If you're inside a bank, speaking with a teller, and they ask you to confirm personal information – that's a safe, expected interaction. You're in the right place, talking to the right people.

The same applies online. If you're using your bank's official app or logged into their secure website, you're inside the Digital Branch – the online equivalent of being at the counter. That's where sensitive activity should happen.

But if you're standing in your front garden, holding a letter that says, "Urgent! You need to pay now. Insert $500 into this envelope and return it to this address!" - that's not a safe context. That letter could have come from anyone. And it might not be from who it says it is.

The same goes for emails and SMS messages. These are not official contexts for sharing personal information, accessing your account, or making payments. They're like someone shouting instructions from the street – not someone inside the branch.

Just like a letter in the post, SMS and Email alerts are notifications – not an invitation to begin a transaction within that context. Leave the SMS or Email environment and carry out the transaction in the appropriate place.

DO ENGAGE

- With an outbound call to a number you have verified
- Within in your Banking App or Website

DON'T ENGAGE

- Within a Text Message
- Within an incoming call
- Within an Email

A Reminder About Trusted Institutions on Digital Mainstreet

Legitimate institutions – like banks, Centrelink, or state services (Service Victoria, Service NSW, Service Tasmania) - operate with integrity. They won't:

- ⊗ Ask for personal or financial information over the phone
- ⊗ Pressure you into acting immediately
- ⊗ Demand payment through unusual or unverified channels

When dealing with these organisations, always verify through:

- ✓ Their official website or mobile app
- ✓ A direct call to their listed contact number
- ✓ A visit to a physical branch or government office

> **IF SOMEONE CLAIMS TO BE FROM A TRUSTED INSTITUTION BUT ASKS YOU TO ACT FAST, PROVIDE SENSITIVE DETAILS, OR PAY IN AN UNFAMILIAR WAY — IT'S MOST LIKELY A SCAM. EITHER WAY, FEEL EMPOWERED TO STOP AND VERIFY.**

As of 2026, most legitimate organisations now ask you to interact through trusted channels – like their official apps, websites, or verified phone numbers. While a few still include links in genuine text messages, it's always safest to switch context: don't tap the link, go directly to the source yourself. Open the app. Visit the website. Call the number you know is real.

That small step puts you back in control. And if you think a company's message could be clearer or safer? Let them know. Your feedback helps make the digital world better for everyone.

Trust: Earned, Not Given Away for Free

A message for the kind-hearted.

You're someone who sees the good in others. You offer trust freely, because that's who you are – and that's a beautiful thing. But in the digital world, scammers take advantage of people's good nature.

This isn't about changing your character. It's about adding a layer of protection that keeps your kindness safe.

The Trust Rule

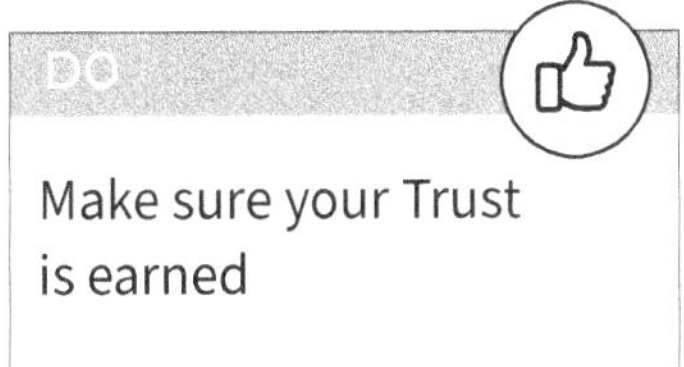

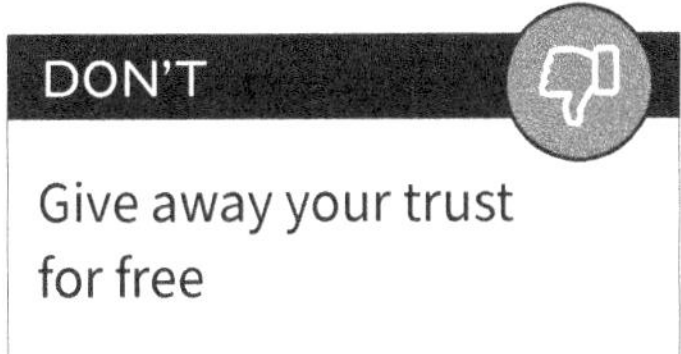

Withholding trust doesn't mean you're suspicious, it means you're practicing digital self-defence.

You lock your front door, not because you suspect your neighbours, but because it's a healthy habit.

Online, the same principle applies.

- You don't assume the worst, you simply wait for proof.
- You don't share sensitive information until you're in a safe context.
- You don't act on urgency until you've verified the source.

This isn't fear. It's confidence. It's knowing that trust comes on merit – not automatically.

Trust is *earned* and you are in control of when and how you give it.

DON'T BE EASILY DISTRACTED

Distraction is one of the scammer's most powerful tools.

Their goal is simple – get you to focus on the actor, so you don't notice their actions.

Whether they're pretending to be a bank representative, a helpful technician, or even a friend using an AI-generated voice, the performance is just a mask. The real giveaway is always in their actions.

Scammers scam. It's what they do. So, if you want to know whether someone is a scammer, don't judge them by how they sound, what they wear, or how convincing their story is. Judge them by their actions. Are they asking for money? Are they asking for personal information? Are they doing this out of context – like in an SMS, email, or unexpected phone call? Are they putting you under pressure? Are they trying to rush you?

This is the habit you need to build: focus on the action, not the actor. When something feels off, pause and ask yourself, "What are they trying to get me to do?" That simple shift in attention can stop a scam before it starts.

Once you build and practice the habit, you'll start spotting suspicious actions – even in subtle scenarios or sophisticated impersonations.

DO FOCUS ON

The Actions

- Are they asking for money?
- Are they asking for you to click on a link?
- Are they asking you to take an action outside of the correct context?
- Are they putting you under time pressures?
- Are they trying to intimidate you?

DON'T FOCUS ON

The Actor

- How believable they appear or sound.
- How impressive their logo is.
- How well worded their message is.
- How much they know about you.
- How legitimate their 'network' may seem, even similar contacts or location to you.

These are examples you may see in all contact methods: SMS, Email, social media, phone calls etc.

THE BASICS OF STOP AND VERIFY ARE FOR EVERYONE

Even experienced users can fall victim to scams, often because of over-confidence, distraction, or simply skipping the basics.

I once visited a vineyard that offered Segway rides through the vines. The instructor gave one simple rule:

"Remember the 4 C's – Cautious, Confident, Cocky, Crash."

I thought of myself as pretty much advanced enough to take off at speed. Sure enough, I got cocky… and crashed.

The lesson? No matter how experienced you are, you need to operate in that middle ground: confident enough to stay safe, but not so confident that you overlook the fundamentals.

Scams work the same way. Even seasoned Technology users can be caught off guard if they fail to verify sources, ignore warning signs, or assume they're too smart to be fooled.

Stop. And Verify.

Whether you're Tech-Terrified or a Tech-Enthusiast, taking a moment to pause, assess, and confirm authenticity is the simplest, most effective way to stay ahead of scams.

So if you're thinking, "I'll never get the skills if even the Tech-Savvy are falling victim," don't worry – that's not true. Because these aren't technical skills, they are human skills. Ones we all need to remember – all of the time.

This chapter is the core of the Skill Tips, and it covers the essential habits that will serve you well in staying safe online. But they're only useful if we practice them and combine them with a deeper understanding of the tools and tactics scammers use.

Don't be a soft target

Scammers are opportunists. They look for the easiest way in – the digital equivalent of an unlocked door or an open window. If you're unprotected, distracted, or unaware, you become a soft target. And that's exactly what they're hoping for.

So far, we've talked about instincts, reflexes, and habits – your internal defences. But now it's time to look at your external defences: the tools and settings that make you harder to trick, harder to impersonate, and harder to hack.

Think of it like home security. You can be alert and cautious, but if your front door has a broken lock and your windows are wide open, you're still vulnerable. The same goes for your digital life.

That's why the final section of The Basics is all about raising your defences – not with complexity, but with two simple, powerful protection measures that dramatically reduce your risk:

Make it harder for scammers to pretend to be you.

PROTECT YOUR DEVICE

Make it harder for scammers to get into your digital home.

These two steps don't just make you safer – they make you more confident. And confidence is your best defence.

PROTECT YOUR IDENTITY

The most important step to protect yourself online is strengthening your identity. That starts with understanding Multi-Factor Authentication (MFA) or Two-Factor Authentication (2FA) - a simple way to prove who you are and stop others from pretending to be you.

I know - 'jargon alert' - but you'll see this everywhere, so it's a good time to demystify what it is. So, let's bring it back to the real world.

Imagine someone knocks on your door claiming to be from your electricity provider. Before letting them in, you'd check:

- A case number they mention – something they know.

- An ID badge – something they have.

- A matching photo – something they are.

These are called factors. MFA and 2FA works by combining two of them – like a password (something you know) and a phone that receives a code (something you have).

It's like double-checking before opening your front door. And it makes impersonation much harder for scammers and cyber criminals.

Because if someone steals your password, they still won't have your phone to approve the login. That second factor stops them in their tracks. So don't see MFA or 2FA as an inconvenience- see it as part of your critical defence.

<table>
<tr><td>DO </td><td>DON'T </td></tr>
<tr><td>

- Use 2FA or MFA for anything that has access to your information or financial transactions

- Stay alert for fake 2FA and MFA prompts

</td><td>

- Rely on a password alone

- Approve 2FA or MFA requests you didn't initiate (Stop and verify! "Was that me?")

</td></tr>
</table>

KEEP YOUR DEVICES UPDATED

Imagine criminals in your neighbourhood had invented a **master key** that could open front doors. You'd update your locks immediately, right?

SCAMMERS OFTEN EXPLOIT WEAKNESSES IN OUTDATED SOFTWARE.

That's exactly what software updates do. When scammers discover new vulnerabilities, developers respond by releasing 'patches' to fix them. These updates are your upgraded locks or improved building materials and act to keep intruders out.

Scammers often exploit weaknesses in outdated software. If your phone, computer, or apps aren't updated, it's like leaving the front door unlocked with a sign that says: "Easy Target."

So, when your device prompts you to update – do it. But make sure you do so from trusted sources.

DO

- Download apps from trusted sources like your phone's app store or official websites.

- Update your phone, computer, and apps when prompted.

- Respond to update instructions from known, verified sources.

DON'T

- Ignore or delay updates -they're often urgent security fixes.

- Download apps from untrusted sources, like links in SMS, emails, or sketchy websites.

- Assume your device is safe just because it's working fine outdated software is a hidden risk.

KEEPING YOUR DEVICES UPDATED ISN'T JUST A TECHNICAL CHORE — IT'S A FORM OF SELF-DEFENCE. EVERY UPDATE IS A QUIET REINFORCEMENT OF YOUR DIGITAL WALLS.

It's easy to ignore those prompts or delay them for later, but each time you do, you leave a door ajar. In a world where threats evolve quickly and quietly, staying current is one of the simplest, smartest ways to protect yourself.

So, treat updates like changing the locks when you know someone's found a way in. It's not just maintenance – it's prevention. And it's one of the most powerful habits you can build.

Chapter Summary

KEY CONCEPT	SUMMARY / ACTION TIP
Reflex Training	Trust your instincts. Pause when something feels off. Your gut reaction is your first line of defence.
STOP and VERIFY	STOP gives you control. VERIFY gives you certainty.
Red Flags	Watch for: Urgency Fear Requests for personal information Payments requests or Too-good-to-be-true offers. Suspicious links. These are signs to STOP.
Outbound Action	Always verify using your own trusted sources. Never trust inbound links, numbers, or instructions.
Inbound Influence	Don't act on unsolicited messages, calls, or emails. They're like strangers at your digital door.
Urgency Trap	Scammers push you to act fast. Pause and disconnect when pressured.
Fear Trap	Scammers use threats to make you panic. Breathe, stop, and regain control.
Context Matters	Safe actions happen in trusted apps or websites. Suspicious actions happen in texts, emails, or calls.
Trust Role	Trust must be earned. Don't give it away freely. Wait for proof before engaging.
Focus on Actions	Judge by what they ask you to do - not how they sound or look. Scammers always want something. Focus on the Actions Don't focus on the Actor
Protect Your Identity	Use MFA/2FA to stop impersonation. It's like locking your digital front door.
Protect Your Device	Keep software updated. Updates are security patches that block scammer access.

CHAPTER 3

COMBAT DRILLS: TRAINING FOR REAL-WORLD SCAM ATTACKS

In Chapter 2, we built the foundations. You learned the core principles that sit underneath almost every scam: how they create pressure, how they exploit emotion, and how your reflexes help you interrupt that moment before damage is done. Those concepts are the groundwork for everything that follows.

But knowing the principles is not the same as being able to use them. A skill only becomes useful when it moves from knowledge into habit. That means training it, practicing it, and reinforcing it until it becomes part of how you naturally respond.

That is the purpose of this chapter. This is where we build the muscle. By working through familiar scenarios in different contexts, you strengthen the reflexes that protect you, and you make sure they do not sit in theory but in action.

Think of this stage as sparring. Repetition is not a weakness. Repetition is how instinct is formed. The more you practice these drills, the earlier you will recognise a setup, and the easier it becomes to break the connection before the scammer gains control.

As we go through the drills, we'll use the Red Flag Checklist as our guide to help guide our Stop and Verify actions.

RED FLAGS CHECKLIST

Use this as a guide to detect potential scams.

RED FLAG	WHAT TO WATCH FOR	CHECK?
Unknown Contact	Unknown number, unsaved contact, unfamiliar person, odd email address	☐ Yes ☐ No
Urgency	"Act now", countdowns, pressure to act before verifying	☐ Yes ☐ No
Fear	Threats of legal action, account suspension, or fines	☐ Yes ☐ No
Context	Requests via SMS/email instead of official app or website	☐ Yes ☐ No
Payment	"Pay now", gift cards, crypto, wire transfers	☐ Yes ☐ No
Links	Login pages, short web links, 'Click here' buttons	☐ Yes ☐ No
Confusion	Baffling you with technical jargon, financial complexity or something you don't understand	☐ Yes ☐ No
Offers	Promises of free money, prizes, exclusive deals	☐ Yes ☐ No
Emotion	Guilt trips, sympathy ploys, help requests	☐ Yes ☐ No
Errors	Poor grammar, typos, incorrect logos, wrong provider name, not a service you use	☐ Yes ☐ No
Unexpected	A charge, email correspondence or interaction you are not expecting	☐ Yes ☐ No

SO MANY SCAMS, YET THE SAME FEW PATTERNS

Scams come in dozens of forms, each trying a different angle and a different story. But for the purpose of these drills, we will focus on only a few examples. These are enough to reveal the playbook that all scams rely on, no matter what shape they take.

This is often the point where people start to feel overwhelmed. Romance scams. Phishing scams. Recruitment scams. Family impersonation scams. SMS scams. Crypto scams. Investment scams. Fake invoice scams. Tech support scams. The list seems endless. It shifts and adapts in ways that make it feel impossible to keep up.

Gone are the days when spotting a typo from a socalled Nigerian Prince was enough to stay safe.

IF YOU HAVE EVER THOUGHT, "HOW AM I SUPPOSED TO PROTECT MYSELF FROM ALL OF THIS?", YOU ARE NOT ALONE. THE VOLUME OF SCAMS, THEIR VARIATIONS AND THE CONSTANT SENSE THAT NEW ONES APPEAR EVERY WEEK CAN MAKE THE PROBLEM FEEL TOO BIG TO MANAGE. THAT FEELING OF BEING OVERWHELMED IS NOT AN ACCIDENT. SCAMMERS RELY ON IT.

But here is the truth. You do not need to memorise every scam type. You only need to recognise the core tactics. Underneath the surface, most scams behave the same way. Their strategies repeat. Their patterns are predictable. They rely on urgency, emotion, pressure and the unexpected. And the skills you are practicing here, built around Stop and Verify and reading the action, apply to them all.

For reference, you can view the full list of scam categories in more detail in Appendix: Scam Types.

Now that you have these foundations, it is time to analyse the patterns using the Red Flags and the tips covered so far. Let's begin with Combat Drill One.

COMBAT DRILL I

FAKE INVOICES AND BILLING

Your phone buzzes – an email from your telco claiming you missed a payment. The logo looks legitimate. The wording feels official. There's a warning saying your account will be suspended unless you act immediately.

accounts@bluet-telecom.com

Urgent: Final Notice Before Disconnection

Date: Mon, 7 July 2025, 9:42 AM

Dear Customer,

We have attempted to contact you regarding your overdue account balance of $129.47. This is your final notice before service disconnection.

To avoid interruption of your mobile and internet services, please make immediate payment via PayID to:

PayID: 0412 678 321
Reference: BlueT-Overdue-12947

Failure to act within 24 hours will result in account suspension and additional reactivation fees.

If you believe this message was sent in error, please contact our billing team via email only: urgent.billing@bluet-telecom.com

Thank you for your prompt attention.

Blue T Telecom Support Team

FAKE INVOICES SCORECARD

RED FLAGS

 **Urgency** — *'Immediate'* payment. Loss within '24 hours'

 **Fear** — Intimidating you with loss of service

 **Payment** — Any payment request should prompt you to verify

 **Unknown Contact**
- Requesting money to be sent
- Email is not from official Telco address

Unexpected — Method of communication or timing unexpected

WHAT *COULD* HAPPEN?

- Lost payment
- Have your data shared as a 'verified target' to other scammers
- Receive more scam messages

WHAT *SHOULD* HAPPEN?

STOP: Red Flags detected. Take control.

VERIFY: Use your information, not theirs to verify.

HOW TO VERIFY?

 Contact the email address you have saved in your mailbox or better still, from the official website

 Check your payments by calling bank, or logging on to official app or website

Learning Points

Remember our Digital Main Street rule – reputable organisations do not demand payments outside their secure apps and official channels. Nor do they use high-pressure urgency tactics designed to push you into acting without thinking.

If this were truly a missed payment, would another few minutes really make a difference? Of course not.

THE SAFEST HABIT IS STILL THE SIMPLEST: STOP AND VERIFY.

This example includes many red flags. This is done for exercise purposes so we can work through each one. But if something has this many obvious red flags in a real scenario, feel free to block the sender and delete the message immediately.

The safest habit is still the simplest: Stop and Verify. If something doesn't feel quite right, slow down and contact them using details you've found yourself. Genuine businesses won't mind you double-checking.

COMBAT DRILL 2

THE FAMILY IMPERSONATION ('HI MUM' SCAM)

A text appears - "Hi Mum, I'm in trouble. My phone is broken. Can you send money?" Your heart races. What if this really is your child? The pressure is intense.

These are classified as 'Family Impersonation Scams' so it may not be 'Mum', but it'll have the same characteristics.

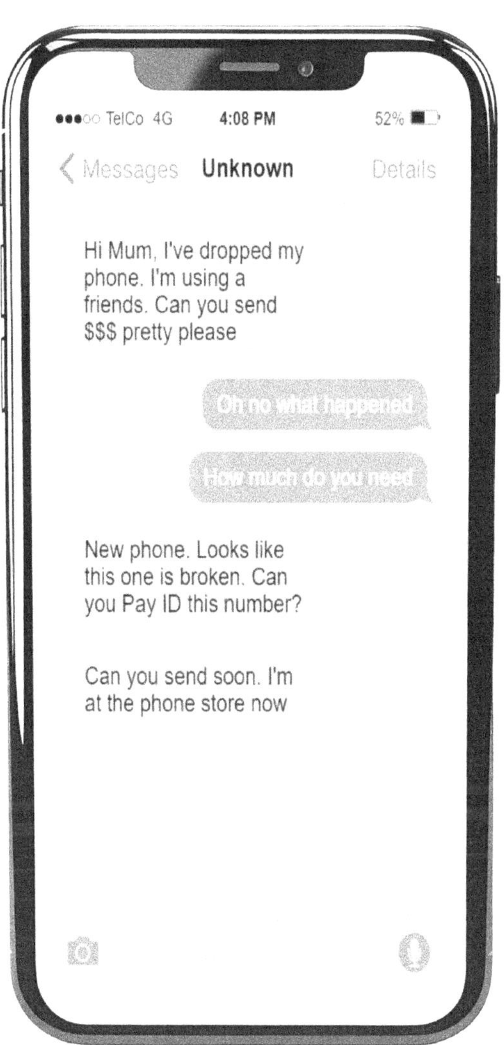
TelCo 4G
4:08 PM
52%
Messages
Unknown
Details
Hi Mum, I've dropped my phone. I'm using a friends. Can you send $$$ pretty please
Oh no what happened
How much do you need
New phone. Looks like this one is broken. Can you Pay ID this number?
Can you send soon. I'm at the phone store now

HI MUM SMS SCORECARD

RED FLAGS

 **Urgency** — *'Immediate'* payment.

Emotion — Emotional blackmail, triggering your instinct to protect your family.

Payment — Requesting money to be sent unfamiliar bank account or Pay ID

Unknown Contact — An unrecognised number

Unexpected — Method of communication or timing unexpected

WHAT *COULD* HAPPEN?

- Lost payment
- Have your data shared as a 'verified target' to other scammers
- Receive more scam messages

WHAT *SHOULD* HAPPEN?

STOP: Resist Urgency Trap and Emotional manipulation.

VERIFY: Use the information you know, not what you've been sent.

HOW TO VERIFY?

 Do not respond. Call your child on the number you have saved in your phone.

 Last resort. Call the number that's texting you...

Learning points

This one is especially tough in practice. Parents instinctively react when their child is in distress. And if the message said, "Hi Gran…", it's even more manipulative, pulling at the heartstrings in a way that makes hesitation feel cruel.

But before we dive into the emotional side of scams, let's focus on the reflex stage – because emotion makes us more vulnerable to deception.

Typically, scammers use this tactic to steal hundreds of dollars using emotion and urgency triggers with the knowledge that families are prepared to pay these relatively low amounts for people they care about.

In 2024, there was a 37.5% increase in Family impersonation scams in Australia. [1] Given this increase, it's worth putting a spotlight on it. Scammers are also including AI voice impersonations which can make these scams harder to detect.

In our family, we've set up a simple safeguard: a shared "safe word" for emergencies. It was agreed upon around the dinner table and has never been shared digitally.

If someone reaches out – like in the "Dear Mum" scam or maybe a strange phone call – we ask for the safe word. If they can't provide it, we know it's a scam.

COMBAT DRILL 3

THE FAKE TECH SUPPORT AMBUSH

Your phone rings. It's a "Microsoft Support" agent telling you your computer has been compromised. They sound authoritative, serious, and concerned. They claim your machine is infected with viruses and they need immediate remote access to fix it.

Fake Support, could be Fake anything in this case. A fake call from a bank claiming a fraudulent attempt to withdraw funds from your account, a Tax Office official telling you to provide extra details relating to your tax return.

They speak with urgency and authority, seriousness and concern. But they are almost always looking to 'help' and need your information and action.

No Caller ID
Decline
Accept

FAKE SUPPORT SCORECARD

RED FLAGS

 **Urgency** — *'Immediate'* action.

 **Confusion** — Using technical jargon and complex terms

 **Unknown Contact** — An unrecognised number or a service you have not requested

 **Unexpected** — Method of communication or timing unexpected

WHAT *COULD* HAPPEN?

 Unauthorised access to your information and accounts

 Using that information, the scammers then cause further harm like taking money or further compromising your

WHAT *SHOULD* HAPPEN?

 **STOP:** Resist Urgency Trap and attempts to confuse you. If you don't understand what's being asked of you, do not act.

 **VERIFY:** Use the information you know, not what you've been sent.

HOW TO VERIFY?

 Contact your provider - if they claim to be from one you have a contract with - using a number you have sourced

Learning points

Fake support scams are the digital equivalent of someone knocking on your front door and trying to force their way inside. Their goal is simple. They want unrestricted access so they can take your money, your information, or control of your device or accounts.

According to Scamwatch, phonebased scams resulted in the highest reported financial losses in 2024, even though more scam attempts occurred through SMS and email. The reason is pressure. When a scammer reaches you live on a phone call, they can inject urgency, fear and confusion in a way written messages cannot. That realtime interaction is how they maximise their influence.

This is why the STOP part of your reflexes matters so much. You always have the power to end the call. You always control the engagement. Hanging up is not rude. It is protective.

Remember that legitimate providers on Digital Main Street do not behave like this. They do not coldcall you about problems with your account. They do not demand remote access or immediate payments. Any number applying pressure can be safely blocked, reported, or ignored.

When in doubt, STOP. Then break contact. Then VERIFY.

They always want you to trust THEIR information. You should only ever trust YOUR information that you have independently verified.

That simple sequence prevents nearly every fake support scam.

COMBAT DRILL 4

THE SOCIAL MEDIA DECEPTION

A friend request appears from someone you sort of recognise – a name you haven't seen in years, maybe an old coworker, a distant cousin, or someone you once met briefly at an event. The profile photo looks familiar enough to feel safe, but something doesn't quite land.

Before you've had time to place them, a message arrives. A warm greeting. A friendly tone. Quick familiarity. They ask how you've been, why you haven't kept in touch. Then, almost immediately, the conversation shifts – a small request for help, a hint of a personal problem, or a vague financial issue they want to "talk to you about."

It feels personal… but also strangely sudden.

You're not sure whether to trust your memory or your instincts.

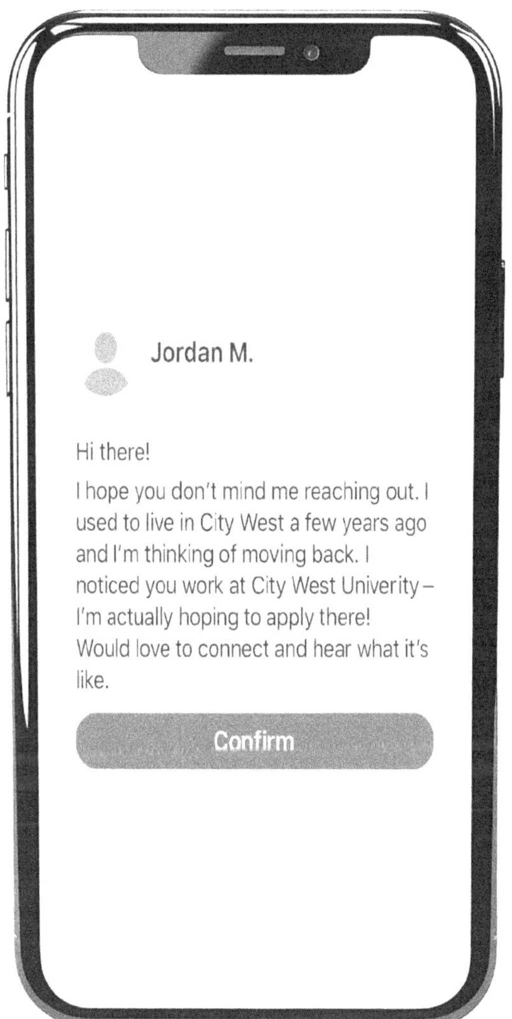
Jordan M.

Hi there!
I hope you don't mind me reaching out. I
used to live in City West a few years ago
and I'm thinking of moving back. I
noticed you work at City West Univerity –
I'm actually hoping to apply there!
Would love to connect and hear what it's
like.

Confirm

SOCIAL MEDIA CONNECTION SCORECARD

RED FLAGS

 **Unknown Contact** — An unrecognised number or a service you have not requested

WHAT *COULD* HAPPEN?

 Develop a connection with you to gain further access to your trust.

 Using your trust against you for financial gain

WHAT *SHOULD* HAPPEN?

 **STOP:** There may be a sense of intrigue at play. Resist it.

 **VERIFY:** Use the information you know, not what you've been sent.

HOW TO VERIFY?

 Research the sender. Look for connection who are connected or search the individual on trusted platforms like LinkedIn or their business website.

Learning Points

The Red Flag at the top of the table is always "Unknown Contact" for a reason. It is the strongest early indicator that you should Stop and Verify. In many social media scam attempts, it may be the only Red Flag you see at the beginning. That alone is enough to take the interaction seriously.

Social media contact is more nuanced because these scams rely less on technology and more on people. There may be no fake links, no obvious demands and no instant push for money. Instead, the deception is built gradually through trust, familiarity and emotional cues. These interactions often appear to come from an individual rather than a business, sometimes even borrowing the reputation of a business or a shared interest to feel legitimate.

Unlike interactions on Digital Main Street, these are unregulated, person-to-person engagements. They rely entirely on emotional manipulation and human psychology. The best defence is to stop the contact before it turns into a conversation. Ending it early prevents the scammer from building rapport or gathering information.

What happens if it does progress? That is where the danger increases. Social media-based scams are slowburn, targeted attempts designed to shape your behaviour over time. This is why an entire chapter of this book is dedicated to these tactics. See Chapter 5, The Targeted Attack, for a full breakdown of how these scams operate and how to defend yourself once the contact has begun.

STICK TO THE PRINCIPLES AND FOLLOW THE FLAGS

Never assume that urgency only appears in phone-based scams, or that fake invoices or missed-payment scams only arrive by email or SMS. The communication channel may change and the language may vary, but the underlying behaviour of the scammer stays the same. That is the positive news. No matter what scenario you face, you can protect yourself by watching for the Red Flags and using the Stop and Verify practices.

Flags, Mixed Flags and No Flags

You should always Stop and Verify, regardless of who is contacting you or what is being requested.

- **Red Flags:** When you practice spotting them, the Red Flags will become familiar and easier to recognise.

- **Mixed Flags:** Even one Red Flag is enough reason to Stop and Verify. If your friends email account is hacked, a scammer could be using it to contact their address book for information or requests for money. A mixture of indicators, such as "Known Contact" combined with an "Unexpected" request or a prompt for "Payment," should always prompt you to pause.

- **No Flags:** Even when everything appears legitimate, it is still worth taking a moment to verify. A parcel notification, for example, may look entirely correct, but checking your Post Office app helps reinforce the habit.

Building the Muscle – Training Your Scam Defence Instincts

Because habits are the key. Turn the knowledge of these principles into practice and develop the habit. Before long, it will become second nature and you will spot potential scams even earlier in their track. But no matter how confident you get, don't get overconfident and crash. Calm consistency will be your friend.

WHERE WE GO NEXT

You might be reading these combat drills and thinking, "This seems a bit basic. These scams are easy to spot in the cold light of day."

That is a fair point. Some scams are simple and obvious when you are calm, wellrested and reading about them in a book. But others go much deeper. They are more nuanced, more personal, and far more devastating. Before we deal with those, it is important to be grounded in the foundational principles and the most common examples. These early drills give you the skills, reflexes and language you will rely on later.

In the next part of the book, we will take a short pause from the Skills Chapters to understand the enemy better. Scammers do not simply appear out of nowhere. They organise, plan, test and refine their approaches.

Understanding how scammers operate is essential. When you learn what motivates them, how they structure their attacks and why certain people or situations are more appealing to them, your defences grow significantly stronger.

So that is where we go next. We will look behind the curtain. We will examine the psychology, motivations and mechanics of the attackers and the industry they have created around scams.

You have the foundations. Now we start studying the opponent.

Chapter Summary

KEY ACTION	TAKEAWAY
Repetition Builds the Habit	Train your reflexes until scam detection becomes second nature.
Scam Patterns Repeat	No matter how different they seem, all scams follow the same core tactics.
Urgency & Action Are Red Flags	Scammers use pressure tactics-analyse what they want you to do.
Ignore the Disguise, Read the Moves	Don't focus on who they claim to be-focus on their demands.
Combat Training - Recognizing Scam Tactics in Action	Regardless of the scenario, all scams follow the same principles. Namely, an impression of someone else, applying trickery or force, to extract your money or information.
Don't get emotional (yet)	Don't give in to emotion (response to pressure, fear, emotional blackmail) When you practice the skills and make them a habit - you can override that emotion.

CHAPTER 4

KNOW YOUR ENEMY: HOW THE SCAMMING WORLD OPERATES

The previous chapters focused on what most people may already know about scams. The foundational self-defence skills, the muscle memory of Stop and Verify, and the combat training to turn awareness into instinct will be fundamental to your armory.

But now, we shift focus to learn about the enemy.

This chapter is less about action, but it is nevertheless crucial. Because knowledge is power. The deeper you understand who your enemy is, how they operate, and why they target victims, the stronger and more prepared you'll be.

SCAMMERS AREN'T VAGUE FIGURES HIDING IN THE SHADOWS. THEY EXIST. THEY OPERATE LIKE A BUSINESS, RECRUITING, TRAINING, AND SCALING DECEPTION LIKE ANY ORGANISED INDUSTRY.

Unlike the previous chapters, this chapter won't be about a table of actionable takeaways. We will take a short break from that to take some time to understand who the scammers are, what they do and why they do it.

This will help set you up for the next actionable takeaways in the chapters that follow by being more informed and: Knowing Your Enemy.

WHO'S REALLY ON THE OTHER END?

A message appears. A notification flashes. A number you don't recognize.

For those who respond, who take the bait – who are they really talking to?

We have all seen the stereotype: the cliché image of someone in a black hoodie, hunched over a laptop with green code cascading down the screen. TV shows, news reports, and stock photos have cemented this image of online crime as something shadowy and mysterious. That perception is not helpful when you are trying to understand what is really happening.

Forget the stereotype. The reality is far more sobering: not lone hackers but deception on an industrial stage.

When your phone rings, your SMS pings, or your social media alert dings with a scammer on the other end, it's not some lone hacker typing away in the dark. It's the scam factory calling.

Cities of Deception

There are many layers to the scam industry. From Solo opportunists who deal in low-level scams, to fraud networks of small teams that operate in an organised manner with shared physical and digital services to run their operations.

The characteristics of these scammers are generally covered Chapters 2 and 3.

But it's important to understand just how large and organised these scam centres are.

Take KK Park in Myanmar. In 2020, it did not exist. By 2025, it had grown into an entire city dedicated to fraud. Built along the Moei

River near the Thailand-Myanmar border, KK Park featured casinos, hotels, restaurants, nightclubs, brothels, supermarkets, residential blocks, and even a hospital. Behind its walls, multi-story buildings operated as call centres, rows of workers at computers under strict supervision, systematically tricking ordinary people into surrendering their life savings.

The rise of KK Park and similar compounds can be traced back to the late 2010s when China began tightening controls on telecom and online fraud. As enforcement ramped up, many syndicates relocated to Southeast Asia, especially Myanmar, where weak governance, corruption, and armed groups control created the perfect conditions for scam factories to thrive.

This relocation turned isolated fraud into a regional industry, now operating at industrial scale.

WHO ARE THEY? HOW DO THEY GET AWAY WITH THIS?

The criminal syndicates that run these scam operations are provided with the perfect conditions to thrive:

- Corrupt governments willing to turn a blind eye in exchange for financial kickbacks.

- Weak law enforcement unable – or unwilling – to dismantle large-scale fraud operations.

- Militias and private security forces protecting scam compounds from outside interference.

- Economic instability, making it easy to recruit desperate workers into scam operations.

In Myanmar, criminal syndicates control territories through partnerships with border guard forces and militias under the Myanmar Army. These groups provide protection for scam centres in exchange for financial benefits.

In Cambodia, scam operations flourished, where Chinese-run casinos and fraud networks operated with little regulation until international pressure forced some crackdowns.

In Laos, scam centres exist in special economic zones, where local elites profit from illegal activities while law enforcement looks the other way.

These governments don't just tolerate scam operations – they benefit from them. The scam industry funnels billions into local economies, creating a financial incentive for officials to ignore the suffering caused by the scam industry.

THE SCAM INDUSTRY — A CRIMINAL ECONOMY

This isn't chaotic crime – it's business. A highly structured, well-funded, and relentless industry that rivals even the drug trade in profitability. Unlike narcotics, however, scammers don't need physical supply chains – there's no product to smuggle across borders. Instead, the scam economy operates like a decentralized marketplace, fuelled by microservices that ensure fraud remains scalable, adaptable, and impossible to dismantle as a single entity.

Every stage of fraud is specialized, with different players contributing to the larger system:

- Target Finders – They collect lists of people likely to fall for scams and sell them online.

- Fake Profile Makers – They create fake accounts, websites, and apps that look real.

- Money Movers – They hide stolen money using shell companies, crypto, and fake invoices.

- Rogue Developers – They build scam software, websites, and apps.

- Scam Service Network – A web of small services that enable identity theft and money laundering.

Like the real economy, the scam industry relies on interconnected micro-services – but instead of logistics and trade, this infrastructure enables financial deception, identity forgery, and digital manipulation.

BECAUSE IT'S SO FRAGMENTED, IT ISN'T ONE ORGANISATION THAT CAN BE SHUT DOWN. EVEN IF ONE SCAM CENTRE COLLAPSES, ITS SERVICES ARE EASILY REPLACED BY OTHERS IN THE ECOSYSTEM. DISRUPTION MAY SLOW FRAUD, BUT IT NEVER STOPS ENTIRELY.

Scammers don't need physical smuggling routes or controlled substances – just an internet connection and a network of criminals willing to sustain the machine.

YOU'RE NOT THE FIRST VICTIM

Many of these scam factories operate in countries like Myanmar, Cambodia, Laos, and the Philippines, where trafficked victims are forced to work as scammers, trained and scripted to manipulate people worldwide. They aren't just "bad actors" - they are victims trapped in a system that thrives on deception and exploitation.

These victims have been scammed themselves. Tricked into travelling from poorer nations in Africa and across Asia, on the promise of a job and upwardly mobile lifestyle, only to be trapped in these scam factories and subjected to inhumane working conditions.

HUNDREDS OF THOUSANDS OF PEOPLE HAVE BEEN TRAFFICKED INTO SCAM CENTRES, OFTEN UNDER FALSE JOB OFFERS.

Human trafficking for scam operations is a serious and growing issue, particularly in Southeast Asia, where criminal networks force victims into online fraud schemes. Estimates suggest that hundreds of thousands of people have been trafficked into scam centres, often under false job offers.

In 2024, the extent of this humanitarian issue is staggering.

- In Myanmar, reports indicate that at least 120,000 people may be held in scam compounds. [4]

- In Cambodia, estimates suggest around 100,000 victims are trapped in similar conditions. [4]

- Other countries, including Laos, the Philippines, and Thailand, also have tens of thousands of trafficked individuals forced into online scams.

Victims endure horrific conditions, including forced labor, torture, debt bondage, and even threats of organ removal. Many are recruited through fake job ads, promising high salaries, only to be coerced into criminal activities upon arrival.

Interpol has targeted these operations, conducting raids across 27 countries, leading to arrests and rescues. However, the problem is expanding globally, with scam centres appearing in Latin America and beyond. And these raids are not harming the scam industry enough.

Victims are recruited from Africa, South Asia, and Southeast Asia, often through social media job ads promising high salaries, free accommodation, and stable employment.

Life Inside a Scam Factory

Once inside, victims endure harsh "working" conditions with unreasonable financial targets. If they fail to scam thousands per week, they are:

- Beaten, subjected to torture, or starved as punishment.

- Moved into harsher roles, including forced prostitution within the compound.

Some victims are even forced into romance scams, posing as real love interests – including video calls with scam victims to make the deception more convincing. [7]

Many governments recognize the severity of scam factories, but their responses vary widely. Poorer nations lack the diplomatic weight to impose international pressure or sanctions on the countries harbouring these operations – some don't even have embassies in key locations, making diplomatic efforts near impossible. [5]

Who's taking action? And why it's not enough

- United Nations & Human Rights Groups – Experts warn that scam factories are a humanitarian crisis, urging governments to act. But without direct intervention, trafficked victims remain trapped. [4]

- Interpol & Law Enforcement – Raids have freed some victims, but scam centres relocate and rebuild fast, making enforcement largely ineffective against the fragmented nature of scam operations.[5],[1]

- Tech Companies – Platforms like Facebook and Telegram attempt to remove fraudulent profiles or job ads, but traffickers adapt instantly, shifting tactics and creating new identities and channels overnight.[1]

They didn't spend millions just to steal a few hundred

The scam industry isn't small-time fraud – it's a global financial powerhouse, siphoning trillions from victims worldwide. Criminal networks don't invest millions into scam factories, fake platforms, and laundering operations just to steal pocket change – they're after life savings, assets, and financial ruin.

The Scale of Scam Profits

- Over $1 Trillion Stolen in 12 Months – A 2024 report revealed that international scammers siphoned over $1.03 trillion globally—a figure that rivals the GDP of some nations. [5]

- Billions Lost Annually in Individual Countries – In Australia alone, scam losses exceed billions per year, with fraud victimisation rates steadily increasing. [8]

- High-Value Targets – Scammers don't just target random individuals – they profile victims, seeking those with can get access to funds easily, and lure them into investment interests.

Once scammers extract money from victims, they can't just spend it directly – they need to launder it, making it appear legitimate before integrating it into the financial system. This process involves multiple layers, using industries that allow large cash flows with minimal scrutiny.

Common Money Laundering Tactics in the Scam Industry

- Gambling & Casinos – Scammers convert stolen funds into chips, play minimal games, then cash out as "winnings", making the money appear legitimate.

- Real Estate Investments – Fraudulent funds are poured into property purchases, often through shell companies, allowing criminals to store wealth in assets while avoiding detection.

- Luxury Goods & Art – High-value items like watches, jewelry, and artwork are bought with scam money, then resold or transferred, disguising the original source of funds.

- Cryptocurrency Laundering – Scammers move funds through crypto exchanges, using privacy coins, mixing services to erase transaction trails.

- Fake Invoices & Shell Companies – Fraudsters create fake businesses, issuing false invoices to justify large transactions, making stolen money appear as legitimate business revenue.

- Trade-Based Laundering – Criminals overvalue or undervalue goods in international trade, using fake transactions to move money across borders without triggering financial scrutiny

WHY THIS IS IMPORTANT TO KNOW?

The conditions inside scam cities are appalling. The treatment of human beings – both the trafficked workers forced into fraud and the victims being deceived – is abhorrent. This isn't just crime – it's a human rights crisis, built on manipulation, coercion, and destruction.

But understanding how this industry works isn't just about exposing its brutality. It's about self-defence. Knowing who they are, how they operate, and the depth of their commitment. Coupled with the knowledge that the industry is extremely organised and profitable, multi-layered and unlikely to be dismantled and run by very dangerous people.

Thankfully, KK Park was raided and largely shut down by Myanmar military forces in October 2025 – however, this was not a permanent victory. The underlying networks quickly adapt, and other scam centres continue to operate. The lessons from KK Park remain critically valuable.

Knowing how low they will stoop, how determined they are to cause pain and how well resourced they are to reach their goals sets us up for the next chapter: The Targeted Attack: When Scams Get Personal.

Chapter Summary

This chapter doesn't replicate the critical work of humanitarian organisations or investigative journalists. Their work is essential and irreplaceable.

Instead, I hope it gives you the awareness you need to strengthen your practical foundation – one you'll rely on as we move into the next chapter, where targeted scams become deeply personal.

DO 	DON'T 
Believe it's real This is happening now - scam cities, trafficked workers, industrial-scale fraud.	**Dismiss it as exaggerated** This isn't hype or paranoia - it's a documented global crisis.
Recognise their cruelty is endless They use torture, coercion, and manipulation without limits.	**Assume they have moral limits** They will exploit anyone, by any means, without remorse.
Understand their scale and structure Scam operations run like businesses with departments and KPIs.	**Think it's just a few bad actors** This is a system, not a handful of rogue individuals.
Expect high investment in fraud Entire cities and platforms are built to sustain deception.	**Assume it's low-effort crime** They spend millions to steal billions - it's strategic and long-term.
See the victims behind the scams Many scammers are trafficked individuals forced into fraud.	**Dehumanize every scammer** Some are victims too. Trapped and coerced into criminal roles.
Use this knowledge as armour Understanding their tactics strengthens your digital self-defence.	**Rely on instinct alone** Gut feelings aren't enough - defence requires informed awareness.

CHAPTER 5

TARGETED ATTACKS: WHEN SCAMS GET PERSONAL

In the last chapter, I laid out the breadth and scale of the scam industry – exposing the low depths scammers are willing to go to.

In this chapter, I will focus on the lowest form of their depravity. In the world of scam deception, there's one phrase that defines the ultimate kill. It's cold, calculated, and ruthless – it's called 'Pig Butchering'.

This scam isn't about taking pennies or dollars. It's a long con, designed to drain victims entirely – financially, emotionally, and psychologically. It's confronting to learn about and devastating to experience.

It is crucial to learn how this type of scam works, why it's so effective, and most importantly – how to defend against it.

THE ULTIMATE CON — UNDERSTANDING 'PIG BUTCHERING'

The term Pig Butchering originates from China, where it describes the process of fattening up a pig – not for the animal's benefit, but for the maximum gain of the butcher. The pig is fed, nurtured, and cared for, only to be slaughtered at the peak of its value.

Side note, If the term makes you angry, if it makes you despise the enemy even more, then perhaps that's exactly the focus needed to tackle this subject.

Scammers apply the same ruthless strategy to their victims. They cultivate trust, build emotional connections, and offer small wins – whether in fake investment schemes or romantic relationships. The victim is slowly conditioned to believe in the scam, investing more time, money, and emotion.

Then comes the final kill, the moment when the scammer pressures the victim to go all in. Not just with their savings, but with everything they can borrow, liquidate, or even steal. The scammer drains them completely, leaving them financially and emotionally devastated.

Some people have trouble believing these types of scams are real. For those – remember from Chapter 4 and who we are dealing with.

PIG BUTCHERING ATTRIBUTES	WHAT YOU NEED TO REMEMBER
They want everything.	Remember how depraved they are.
They are invested for the long run.	Remember how resourced they are.

WARNING: This seems like conflicting advice!

In the early chapters, we focused on identifying scam emails, text messages, and fraudulent contacts. The key lesson was that impersonators are highly skilled, but the real giveaway was in the action they requested, like clicking a link, making a payment, or responding within a hurried timeframe.

Pig Butchering is different. At first, they won't ask for any action at all. Instead, they will seek a connection. There's no urgency, not yet. They take their time, carefully fostering trust until you are invested emotionally, psychologically, and financially.

This shift is critical to recognize:

- Traditional scams rely on urgency and quick deception and rushed action.

- Pig Butchering scams play the long game, using connection-first deception before the real exploitation begins.

The rules don't change, we will always Stop and Verify. The principles of questioning urgency, scrutinizing requests, and rejecting blind trust remain critical.

But now, we must expand what signals we need to be aware of.

TECHNOLOGY — A WEAPON FOR CONNECTION AND ISOLATION

Technology has revolutionised how we connect. It allows us to communicate instantly, build relationships across distances, and access opportunities that were once impossible. The digital world is not something to fear – it's something to navigate wisely.

But 'connection' is only half the story. The other side of technology – the one scammers prey on – is isolation.

A phone is always within reach, always ready to receive messages, notifications, and companionship – 24/7, no matter where you are or how alone you feel. This is why Pig Butchering works. It isn't about quick tricks or urgent fraud; it's about taking advantage of the moments when people feel most alone.

Scammers understand this. They offer comfort, validation, and hope, carefully crafting a fraudulent connection that feels real. This is how they trap victims over time, ensuring that when the moment comes, they'll believe the lie completely.

Weaponising Human Nature – Exploiting Vulnerabilities

Pig Butchering scams work for one reason: they identify your weakness, then exploit it. These aren't generic scams with mass-produced messages, they are tailored specifically to you, your identity, and your emotional triggers.

Scammers often imitate personal traits, values, interests, or demographic markers to build rapid trust, leveraging the psychological principle known as the Similarity-Attraction Effect [1].

Why? Because familiarity breeds trust.

- We connect with people who remind us of ourselves – it validates our experiences and reinforces our beliefs.

- We trust people we perceive as similar – even if that similarity is completely manufactured.

- Scammers mirror you – they study your interests, vulnerabilities, and emotional needs, then reflect them back at you to build trust.

This isn't accidental – it's deliberate psychological manipulation. The scammer becomes your ideal connection, ensuring that when the time comes, you're fully invested and fully convinced.

A PUNCH TO THE HEART — THE BRUTALITY OF ROMANCE SCAMS

The Illusion of a Perfect Connection

Let's see how this approach unfolds in the real world.

Imagine receiving a social media message from an attractive woman overseas.

She's reaching out as she's applying for a job at the university you work at and is eager to return to Australia once her visa is approved. She asks about the culture of the university, whether it's still the same, apologizing for being forward but hoping for insight from past faculty names she remembers. She remembers you being well regarded.

You check her profile. Her work history includes companies in your city and is a graduate of one of the faculties. Like you, she loves bushwalking, with photos of hikes you've personally been on.

At first, the conversation feels harmless, professional. But soon, it shifts.

She casually mentions your shared interest in bushwalking. The conversation deepens.

She shares. You share-and before long, there's a hint of a spark.

Why It Feels Real

Never mind that your public profile shows:

- You work at the university.

- You live in the city.

- You share photos of bushwalking.

- Your social media suggests you're single, middle-aged, and financially stable.

This scam isn't random-it is crafted specifically for you.

It feels real because it is designed to feel real. Every detail has been carefully curated from public information, ensuring it resonates on an emotional level.

The Trap – Confirmation Bias

Your interest is triggered. You want to believe it's genuine.

- Confirmation bias kicks in. You ignore doubts, focusing only on what seems true.

- Your 'verify' attempt is weak. She has a few shared connections, a believable work history, and a credible reason to reach out.

- The connection is made. And that's the moment the scam truly begins.

From Professional Courtesy to Emotional Investment

The progression feels natural, effortless-but it is entirely by design.

At first, she was just seeking guidance-a polite inquiry about the university and the job market. You offered pointers, genuinely wanting to help. Now, her application is under-way, and you're emotionally invested in her success.

She shares personal struggles-her mother's illness, her commitment to saving, working, and making a better life. You admire her determination. She's not just a job applicant now-she's someone you respect, someone you care about.

Then, the shift happens.

- The conversation moves to WhatsApp, more personal, more private.

- You message multiple times a day, sharing stories, ideas, moments.

- Before long, she's the first message you wake up to-the last before bed. Now messages are signed with an 'x' at the end to signify a kiss.

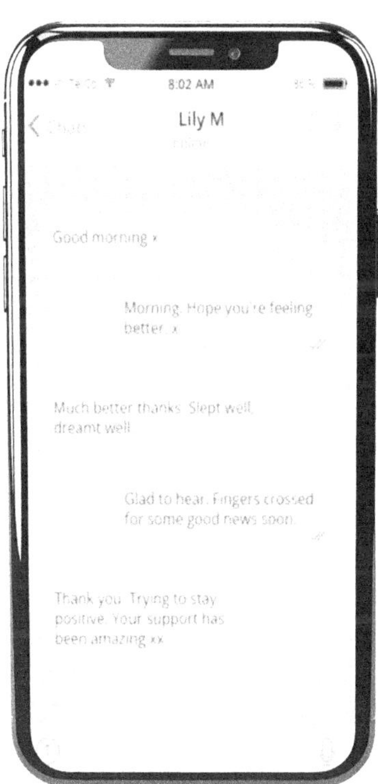

At this point, it doesn't feel like a scam at all, because it isn't just about money yet. It's about trust, familiarity and closeness. The scammer has built a routine, an emotional dependency.

This is Pig Butchering's master play. It never feels transactional, because it's not just a scam, it's a psychological conditioning process. You aren't being pressured. You're choosing to engage and choosing to believe.

The Switch – From Trust to Financial Manipulation

And when the financial 'ask' finally comes, it feels like a natural extension of everything you've already shared.

She takes your advice. She's determined to help her mother financially and return to Australia. A family friend introduced her to a regional cryptocurrency investment-she was unsure at first, but with small amounts, she achieved early profits.

Now, this opportunity could fast – track her return.

You're unsure. But you've been meaning to suggest a call anyway – this moment feels like the perfect chance to deepen your connection. You arrange a video call, finally speaking face-to-face.

The doubts you have vanish.

Seeing her, hearing her voice, watching her speak with sincerity – it all feels legitimate. You've never heard of a Scam Factory, so why would you suspect that some of the women behind these fake profiles actually exist in real life, trained to make deal-closing calls? Or that these scam factories run thousands of fake profiles, warming them up and aging them for believability by embedding them in networks and liking mundane content.

So you believe the person you've connected with.

She shares payout details, screenshots from the trading platform, and offers to help.

You agree. You log into the investment app (it's fake) and deposit a few hundred dollars.

And unbelievably, it doubles within a week.

This actually works.

Or at least, that's exactly what they want you to believe.

The Final Phase – The Butchering

The dopamine rush takes hold. The investment keeps growing, the returns are consistent, and the timeline to be united is getting shorter.

Why stop now?

Maybe this is love. Maybe this is fate. Maybe this is the second chance that was never supposed to happen. These past few months have turned despair into hope and promise.

- More money – why not? Enough for a house, one you could share together.

- More deposits – it's working, it's paying off.

- Superannuation, savings, equity – you name it.

This isn't reckless anymore, it's a plan, a future, a dream that feels within reach.

The investment is committed. The money is transferred. We've reached the slaughter.

The line goes cold.

There was no visa application.

There will be no relationship.

There is only financial ruin, emotional devastation, and betrayal so personal that the damage is irreversible.

The pig has been butchered.

Still doesn't sound believable?

This scenario is completely fictional – but not far-fetched. It's inspired by real stories, and while the details may be imagined, the tactics are very real. If your first thought is, "This doesn't apply to me – I'm retired," or "I don't work in a university," then it's worth pausing.

That reaction is understandable, but it risks overlooking the deeper point: scams are designed to feel personal. The most effective attacks are tailored to your life, your habits, your location, and your interests and it's impossible to put one for every reader of this book. But, whether you're working, retired, studying, or somewhere in between, the scam won't look like this exact example – it will be crafted to feel familiar and relevant to you. And when it does, it won't feel wrong – it will feel completely right.

This is why the chapter 'Know Your Enemy' is so important. Too many people think that targeted attacks are not plausible. So, if you find yourself asking:

"Who would target little old me?"

"Who would waste months pretending to get to know me?",

"My connection is genuine, I spoke to them on a video call."

Then remember who your enemy is, how well they are resourced and how cruel they are prepared to be.

THE DARK ALLEY — WHEN DESPERATION BECOMES EXPLOITATION

Desperation is the common thread with romance scams and financial fraud. When hope is fading, when stability feels just out of reach, people look for a way out; a promise, a shortcut, a miracle.

And in the digital age, there's always something or someone ready to offer exactly that.

A romance scam doesn't begin with deception. It begins with loneliness. A financial scam doesn't start with fraud, it begins with desperation. Even when the red flags are blinking, waving, screaming, some victims choose to ignore them, because what if this time it's real?

That's why some willingly walk down the Digital Dark Alleys, even when everything tells them not to.

To understand it, just look at the similarities with online gambling. They share one key thing in common; the smartphone in your pocket, available 24/7, designed to be there for you when others aren't. It's a gateway to connection or destruction, tailored just for you, feeding your emotions, shaping your decisions.

Scammers know this. They design experiences that feel personal, engaging, rewarding, keeping you hooked with dopamine-fuelled highs until the moment of collapse.

And when that collapse comes, it doesn't just take your money, it takes your trust, your confidence, and sometimes, even your sense of self.

PIG BUTCHERING — QUICK IDENTIFIERS

1. Identifying Profile Characteristics

- Attractive personas and stolen profile images used to appear trustworthy or desirable.

- Highly curated online presence, often showing luxury lifestyles or financial "success."

- Hidden identity, always avoiding phone or video calls to prevent exposure.

2. Identifying Engagement Behaviours

- "Wrong number" style cold messages that feel accidental but are actually scripted openings.

- Lovebombing and emotional pacing, providing attention, praise, or companionship during grooming.

- Longform connection building, investing weeks or months in conversation before mentioning money.

- Gradual introduction of investment talk, often framed as a friendly tip, secret method, or private opportunity.

3. Identifying Scam Actions

- Moving you from Social Media messaging service to direct messaging services like WhatsApp.

- Push victims toward fake investment websites or apps that mimic real trading platforms.

- Encourage initial small investments and allow early withdrawals to demonstrate false "wins."

- Shift victims to cryptobased platforms, making tracing and recovery difficult.

- Rapid escalation of deposits, using urgency, emotional pressure, or manufactured opportunities.

Scams generally fall into two forms. The first and most common are general scams. These are broad, fast-moving attempts designed to snatch quick wins by exploiting fear, urgency and the everyday technical gaps most people have. They do not need to know much about you. They rely on catching you off guard, pushing you into reacting quickly, and overwhelming your attention long enough for you to click, comply or confirm.

The second form is very different. Pig butchering scams are targeted, patient and deeply manipulative. Instead of rushing you, they build a relationship over time, grooming their targets through trust, emotional connection and perceived opportunity. Their goal is maximum financial extraction, and the damage they cause is both financial and personal.

DESPITE THEIR DIFFERENCES, BOTH TYPES COLLAPSE WHEN YOU APPLY THE SAME DISCIPLINE: STOP AND VERIFY.

Despite their differences, both types collapse when you apply the same discipline: Stop and Verify. Slowing down, stepping back and independently checking what you are being told remains the single most effective defence, whether the scammer is trying to frighten you into acting fast or trying to befriend you into letting your guard down.

Comparison Summary: General Scams vs Pig Butchering Scams

CHARACTERISTIC	GENERAL SCAM	PIG BUTCHERING SCAM
Impersonator	Usually an Organisation	Usually an Individual
Action the scammer wants	To take an action, like click a link or make a payment	To make a connection, like accept a friend request and build trust
Urgency	Immediate	Longer Term
Preys on	Low Technical Skills Fear of Authority	Loneliness, Promise of Love Financial Hardship
Goal of the Scammer	Quick Payout	Extract Maximum Payout
Typical Contact Methods	SMS, Email, Websites, Phone calls	Social Media, Dating Apps Phone calls, Messaging Apps

Authorities confirm these scams are actively targeting people across the country, with Australian losses running into the billions – particularly in investment and romance scams, which cause the greatest financial harm [2].

According to the Global Anti-Scam Alliance (GASA), Australia now ranks 3rd in the world for average scam losses per affected person, making it a significant hotspot for this type of attack [3].

But the financial toll is only one part of the story. The emotional damage is profound. Victims often experience psychological distress, shame, trauma, and grief comparable to losing a partner.

When you add the suffering of trafficked workers forced to operate scam factories, the human cost becomes impossible to quantify.

Strengthening our defences is essential – not just to prevent financial loss, but to protect our wellbeing. Recovering from a scam isn't only about your devices or your bank account; it's about rebuilding your confidence, your trust, and your sense of self.

In the next chapter, we widen the lens and explore how communities, institutions, and everyday organisations can strengthen those defences even further.

Chapter Summary

This chapter showed how targeted attacks operate – calculated, patient, and deeply personal. But they're not mysterious.

WHEN YOU COMBINE WHAT YOU KNOW ABOUT COMMON SCAMS, SCAM PRINCIPLES, AND THE WIDER SCAM INDUSTRY, THEIR TACTICS BECOME CLEAR AND PREDICTABLE.

From here, we'll use that understanding to build stronger, everyday defences that protect you from even the most deliberate manipulation.

KEY CONCEPT	ACTION TIP
Long Game Grooming	Scammers invest time to build trust before making any financial request.
Connection First Attacks	Unlike fast scams, the goal here is emotional connection, routine, and dependency.
Psychological Mirroring	They reflect your interests, values, and vulnerabilities to feel familiar and safe.
Isolation & Dependency	They become a daily emotional presence, creating a sense of closeness and reliance.
Direct messaging switch	Watch out for a key red flag: Inviting you to switch from Social Media's messaging service to WhatsApp or other direct messaging platform. Social Media companies regularly shutdown scamming accounts. Scammers try to stay one step ahead by moving you to direct messaging.
Engineered Vulnerability	They target loneliness, financial stress, or emotional need - whatever works.
Scammers Intent	The objective is total financial extraction savings, loans, superannuation, everything.
Same Defence Principles	Stop and Verify still works - but now look for connection based red flags, not urgency.

CHAPTER 6

COMMUNITY DEFENCE

COMMUNITY DEFENCE IS SELF-DEFENCE PLUS

Technology has supercharged the scam industry. It hasn't just made scams easier to launch – it's made them faster, more convincing, and harder to detect. That's why this book began from a technologist's perspective: understanding the tools scammers use is essential for defence.

But technology alone can't solve this problem. Prevention isn't just about the skill tips and combat drills we've practiced in earlier chapters – it also depends on financial awareness and psychological resilience. As I've stated before: scammers don't hack systems, they hack people.

This chapter is about the power of community in both prevention and response. Because while scams are designed to isolate people, 'connection' is the antidote. Not technology connection: human connection.

SCAMMERS THRIVE ON SEPARATING VICTIMS, BREAKING THEIR CONFIDENCE, AND ENSURING THEY DON'T SEEK HELP. BUT AS A COLLECTIVE, WE CAN BE A POWERFUL FORCE AGAINST SCAM INDUSTRIES.

Together, we can share knowledge, spot warning signs, and support those who've been targeted.

GLOBAL, NATIONAL AND LOCAL DEFENCE

Global

Across the world, international organisations, researchers, and trusted industry groups work together to understand global scam trends, track criminal networks, and develop cross-border solutions.

They do this by:

- Running global summits and virtual meetups
- Publishing global scam research
- Sharing realtime threat intelligence
- Bringing governments, tech giants, consumer organisations, and law enforcement together

These include organisations such as:

- Global Anti-Scam Alliance (GASA)
- Scam Adviser (global scam detection and intelligence)
- Interpol & Europol cybercrime divisions
- International consumer protection bodies
- Large technology companies researching online fraud trends

These groups form the outer defence layer – monitoring global patterns so countries like Australia can respond quickly.

Everyday People don't tend to interact with these organisations, but they are a crucial part of the global fight against scams.

National

Across Australia, government agencies, financial institutions, telecommunications providers, and major community organisations form a national response against scams. They translate global intelligence into real world action, protect consumers, and provide the systems everyday people rely on when something goes wrong. They do this by:

- Issuing warnings and public alerts about emerging scam trends
- Providing secure ways to report scams and cybercrimes
- Helping people recover financially or reclaim stolen accounts
- Investigating fraud and coordinating law enforcement action
- Building national education campaigns and safety standards
- Strengthening protections across banking, telco, postal, government and commercial sectors

This includes organisations such as:

- National AntiScam Centre (Scamwatch)
- ReportCyber (ACSC)
- Banks and financial institutions
- Telecommunications providers
- Australia Post and major delivery networks
- National digital inclusion organisations
- Consumer protection regulators and ombuds services

These groups form the national defence layer – the systems and structures everyday Australians interact with directly when they need help. They provide the practical tools, the reporting pathways, and the recovery support that make scam safety accessible to all.

Everyday People engage with this layer regularly, even if they don't realise it: through their bank, telco, government services, or scam education campaigns. It is the backbone of Australia's ability to repel, track, and disrupt scams at scale.

Local

Local organisations sit closest to the community. They provide person-to-person guidance, trusted help, and practical digital support. These are the places where everyday people can walk in, ask questions, feel heard, and get assistance without judgement. They do this by:

- Offering digital literacy help and one-on-one support
- Delivering scam awareness sessions and safe technology training
- Helping people understand official advice and take next steps
- Connecting vulnerable individuals to specialised services
- Providing community spaces for learning, recovery and support

This includes organisations such as:

- Public libraries
- Community centres and neighbourhood houses
- Local not-for-profits
- Digital mentor programs (e.g., through Good Things Foundation Australia)
- Local councils and community safety programs
- State and territory consumer protection agencies

These groups form the local defence layer – practical, nearby support that helps people build confidence, improve their digital skills, and stay connected.

Everyday People interact with this layer often: at the library, at a community event, in local workshops, or through trusted community networks. It is where global and national efforts become real world help, delivered by people who understand the needs of their local community.

Our own Communities

Beyond the global organisations protecting people at scale, and the national and local services supporting Australians every day, lies the most personal and powerful layer of Community Defence: the community you have already built around you.

This includes the people you trust most – the ones who know you, support you, and are willing to listen when something doesn't feel right.

Your own community includes:

- Family
- Friends
- Neighbours
- Colleagues and teammates
- Sporting clubs and volunteer groups
- Church groups and faith communities
- Book clubs and hobby groups
- Any network where you feel safe asking, "Does this look right to you?"

These people form your inner defence layer, acting as your early warning system.

They help you:

- Sense when something feels off
- Get a second opinion before acting
- Talk through a suspicious message or call
- Stay grounded when emotions are high
- Recover emotionally after an incident

Scammers succeed when people feel isolated. They fail when people feel supported.

Your personal community is not only a source of strength it's one of the most effective tools in digital selfdefence.

OUR ROLE IN COMMUNITY SUPPORT

All of these communities can provide support to you directly or indirectly, before or after a scam event. However, you are also a supporting member of your own community and have a role to play in community defence.

You can contribute a supporting role in 3 ways.

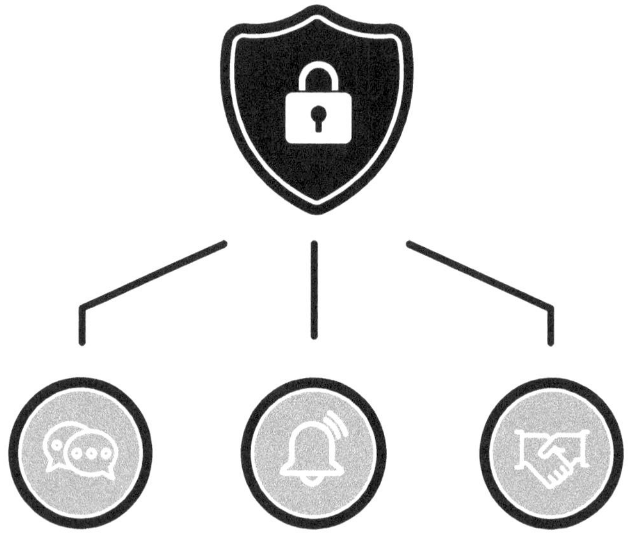

BE OPEN

Share insights and experiences, spreading scam awareness before deception takes hold.

BE ALERT

Watch for signals that someone may be vulnerable or already trapped in a scam.

BE THERE

Support victims without judgment, creating a safe space for recovery.

Be Open – Share With Others

Awareness doesn't require expertise – just conversation.

A single conversation can spark understanding and prevent deception. The more we share what we know, the less likely scams will succeed.

- Discuss scams you've seen, whether online or in your personal life.
- Report fraud attempts to Scamwatch or relevant authorities.
- Encourage open discussions in workplaces, friend groups, and communities.

SCAMMERS COUNT ON SECRECY. THEY HOPE VICTIMS FEEL TOO EMBARRASSED TO SPEAK UP. WHEN WE REMOVE THE STIGMA AND NORMALISE SCAM AWARENESS, WE WEAKEN THEIR INFLUENCE.

Be wary not to scare people when sharing. The intent of this book is to empower people with resilience and make them aware that Scam Self-Defence is achievable and worthwhile.

Be Alert – Watch for Signs in Others

You may not always see the signs – sometimes they remain hidden until it's too late. But by understanding how scammers operate, you can recognize key red flags:

- A friend suddenly talking about investments in an unusual way or cryptocurrency for the first time.
- Someone asking for personal loans with urgency.
- A loved one becoming withdrawn or secretive about online interactions.

IF YOU NOTICE CHANGES THAT RAISE CONCERN, START A CONVERSATION — A SIMPLE DISCUSSION ABOUT SCAMS COULD PREVENT A WORLD OF PAIN.

It only takes one conversation to break someone's isolation.

Be There –
Support Without Judgment

People will fall for scams, despite everyone's best efforts – and they will need help.

- Be someone they can turn to without fear of judgment.

- Offer emotional support – victims often feel ashamed and need reassurance.

- Direct them to trusted resources for financial, legal, and technological assistance.

Victims often fail to report scams out of shame, guilt and a fear of judgement. This lowers the awareness within their community and means we have a lower understanding of the true magnitude of the impact of scams.

However, we don't need victims to broadcast their information publicly, but within trusted and supported circles this could be crucial.

RECOVERY STARTS WITH CONNECTION. VICTIMS NEED REAL HUMAN SUPPORT TO REBUILD CONFIDENCE AND NAVIGATE THE PATH FORWARD.

ONLINE PROBLEMS ›
OFFLINE SOLUTIONS

The strongest defence against scams isn't found in algorithms –
it's found in people, connection, and shared knowledge.

SCAMMERS EXPLOIT HUMAN VULNERABILITIES, NOT JUST TECHNOLOGICAL GAPS.

While awareness of digital threats is important, the most devastating harm occurs when they create fake emotional connections
with precision and intent.

- Loneliness and financial desperation can't be fixed in this book
  – but connection can help rebuild trust.

- In-person conversations provide deeper support – whether
  through friends, family, or professional services.

RESOURCES

Further Education and Awareness – When You Want to Learn More

Knowledge is power. If you'd like to build your digital confidence or help others do the same, these national programs offer free resources and community support:

Be Connected – Free online courses and guides to help older Australians build digital skills

https://beconnected.esafety.gov.au

Good Things Foundation Australia – A network of community partners delivering digital literacy programs and support across the country

https://goodthingsaustralia.org

Emotional Support – When You Feel Distressed, Anxious, or Alone

Scams don't just cause financial harm — they cause emotional harm too. If someone feels isolated, ashamed, overwhelmed, or unsure what to do next, these national services offer confidential support:

Lifeline – 24/7 crisis support and suicide prevention
13 11 14

Beyond Blue – Support for anxiety, depression, loneliness
1300 22 4636

FriendLine – A free chat line for people feeling isolated, lonely, or needing someone to talk to
1800 424 287

These frontline services ensure no one is left behind. Local services may also be available, but these national providers are the best place to start.

Financial Defence & Recovery – When Your Money or Accounts Are at Risk

Your Bank or Credit Union

Your first call when:

- Your financial details are stolen

- You see suspicious transactions

- You need to *Stop and Verify* if a message is legitimate

Most banks now have AntiScam Teams and regularly publish:

- Current scam warnings

- Customer alerts

- Safeverification practices

VERIFICATION TIP — Use Official Websites

Many commercial organisations (banks, telcos, delivery companies, utilities) publish Scam Alerts on their websites.

Never use links or phone numbers inside a suspicious message.

Instead:

- Go directly to the organisation's official website or app

- Search for "Scam Alerts"

- Compare the message you received

Call only the number listed on the official website or the number printed on your bank card

If it's real, it will be listed officially. If not, treat it as a scam.

National Debt Helpline

https://ndh.org.au

For free financial counselling and guidance.
1800 007 007

Australian Taxation Office (ATO)

https://www.ato.gov.au

Provides:

- Scam identification
- Verification steps
- Guidance on responding to impersonation scams

Information, Identity & Account Defence – When Your Data Has Been Breached

IDCARE

https://www.idcare.org/individuals
1800 595 160

Australia's national identity and cyber support service.

Helps with:

- Identity theft
- Account compromise
- Personal data breaches

Email or Service Provider

If your email or social media is compromised:

- Change your password immediately
- Enable MultiFactor Authentication
- Review recovery options
- Remove suspicious app access

ReportCyber (ACSC)

https://www.cyber.gov.au/report-and-recover/recover-from/hacking

For cybercrimes such as:

- Online fraud
- Identity theft involving digital access
- Blackmail and extortion
- Hacking
- Malware incidents

Reports are automatically forwarded to the correct police agency.

Government & Consumer Protection – How Australia Fights Scams

Scamwatch (National AntiScam Centre)

https://www.scamwatch.gov.au/report-a-scam

Created to:

- Help Australians recognise, avoid, and report scams
- Track scam trends using community reports
- Warn the public about emerging threats
- Inform national prevention campaigns
- Report scams here to help protect others

State & Territory Consumer Protection Agencies

For issues involving businesses, contracts, fake invoices, refunds, or consumer rights disputes.

STATE / TERRITORY	AGENCY
Tasmania	CBOS - Consumer, Building and Occupational Services
NSW	NSW Fair Trading
Victoria	Consumer Affairs Victoria
Queensland	Office of Fair Trading (OFT)
Western Australia	Consumer Protection WA
South Australia	Consumer and Business Services (CBS)
Northern Territory	NT Consumer Affairs
ACT	Access Canberra - Fair Trading

When to Contact Consumer Authorities

- A business has misled you
- You were scammed via a trader, invoice, or purchase
- A product or service was misrepresented
- You need consumer rights advice or dispute resolution

Community Support – Trusted Local, Grassroots Help

Libraries

Often provide:

- Scam awareness sessions
- Digital literacy help
- Support with emails, devices, passwords
- A safe place to ask questions

Good Things Foundation Australia

https://www.goodthingsfoundation.org.au/

Leader in digital inclusion programs across the country.

Supports:

- Community centres
- Digital mentors
- Older Australians

Help Starts Here

https://www.helpstartshere.org.au

A national directory for local services, community programs, and support pathways.

Local Community Centres & Not-For-Profits

Provide:

- Oneonone help
- Learning workshops
- Peer support
- Trusted digital literacy assistance

Quick Reference: Who helps with what?

TYPE OF SUPPORT	WHO HELPS	WHEN TO CONTACT
Educational Support	Be Connected, Good Things Foundation, Service Providers	Learning about scams and protection, maintaining awareness of scams
Emotional Support	Lifeline, Beyond Blue, FriendLine	Feeling distressed, overwhelmed, ashamed, or isolated
Financial Defence	Banks, ATO, National Debt Helpline	Money stolen, suspicious charges, financial confusion
Identity & Info Defence	IDCARE, Email/Service Providers, ReportCyber	Identity theft, hacked accounts, data breaches
Consumer Protection	CBOS (Tas) + all state consumer affairs agencies	Scams involving businesses, invoices, contracts
Scam Prevention & Reporting	Scamwatch	Scam attempts, suspicious messages, community alerts
Community Support	Libraries, Good Things Foundation, community centres	Digital literacy help, scam education, confidence building
Verification	Official company websites & apps	To check if a message, call, or email is legitimate

THE FINAL WORD — THE POWER OF SCAM AWARENESS & SELF-DEFENCE

You are no longer an easy target. You are prepared, informed, and empowered.

This handbook shows you how scams work, why they succeed, and most importantly – how to fight back. You now understand the psychological tricks, the digital manipulation tactics, and the real-world scam networks that operate behind the scenes.

But knowledge alone isn't enough.

Awareness must turn into action. Vigilance must become habit. Protection must be continuous.

Use and practice the skill tips, the practice drills and quick references in the Appendix. Always look out for the red flags and build the habit of Stopping and Verifying until it is a natural part of your life.

Scam Defence Is a Lifelong Practice

Just like our security at home, no matter how strong the walls, windows, and locks are, no matter what cameras and alarms we install, we are the ones who still make choices every time we open the front door.

Now that you understand your online neighborhood, you can spot the safe paths to travel and recognize the warning signs along the way. You know who to trust and what actions are trustworthy. You know when and where it's safe to share information, and how to protect your finances when it matters most.

You're ready to make those choices with confidence.

You're ready to open the front door.

Chapter Summary

KEY IDEA	WHAT YOU SHOULD TAKE AWAY
Many layers of support	■ Global → National → Local → Personal ■ Different layers protect you in different ways ■ You're never facing scams alone
Know where to go for help	■ Clear pathways for reporting, recovery, and advice ■ The same places you 'Verify' information can be used for awareness and guidance
Human support, not Tech Support	■ People protect people ■ Everyone can be a recipient or a provider of support ■ Reporting scams can help the next potential victim
Your role in Community Defence	■ Be Open - talk about scams ■ Be Alert - watch for signs ■ Be There - support without judgement

APPENDIX

RED FLAGS CHECKLIST

Use this as a guide to detect potential scams.

RED FLAG	WHAT TO WATCH FOR	CHECK?
Unknown Contact	Unknown number, unsaved contact, unfamiliar person, odd email address	☐ Yes ☐ No
Urgency	"Act now", countdowns, pressure to act before verifying	☐ Yes ☐ No
Fear	Threats of legal action, account suspension, or fines	☐ Yes ☐ No
Context	Requests via SMS/email instead of official app or website	☐ Yes ☐ No
Payment	"Pay now", gift cards, crypto, wire transfers	☐ Yes ☐ No
Links	Login pages, short web links, 'Click here' buttons	☐ Yes ☐ No
Confusion	Baffling you with technical jargon, financial complexity or something you don't understand	☐ Yes ☐ No
Offers	Promises of free money, prizes, exclusive deals	☐ Yes ☐ No
Emotion	Guilt trips, sympathy ploys, help requests	☐ Yes ☐ No
Errors	Poor grammar, typos, incorrect logos, wrong provider name, not a service you use	☐ Yes ☐ No
Unexpected	A charge, email correspondence or interaction you are not expecting	☐ Yes ☐ No

SCAM TYPES

SCAM TYPE	DESCRIPTION	COMMON CONTACT METHOD
Phishing	Fake emails or messages designed to steal login info or personal data.	Email, Fake Websites
Social Engineering	The use of information about you, gathered through social media and other public sources, then used against you to manipulate trust or commit threats - tricking victims into revealing sensitive details or taking risky actions.	Any interaction (calls, messages, social media)
Vishing (Voice Phishing)	Scammers posing as banks or officials to pressure victims into revealing sensitive info.	Phone Calls
Smishing (SMS Phishing)	Fraudulent texts with malicious links or urgent warnings.	Text Messages
Investment Scams	Fake opportunities promising huge returns to lure victims into sending money.	Social Media, Cold Calls
Romance Scams	Scammers build trust in online relationships before requesting money.	Dating Apps, Messaging
Threat & Extortion Scams	Scammers pretend to be law enforcement or tax agents, threatening legal action.	Calls, Emails
Impersonation Scams	Fraudsters mimic trusted businesses, friends, or family using fake credentials.	Emails, Calls, Social Media
Fake Online Stores	Scammers set up fraudulent shopping sites, selling non-existent or low-quality products.	Websites, Ads
Job and Employment Scams	Fake job offers that ask for upfront fees or personal information.	Email, Job Sites

TECHNOLOGY PERSONA I: TECH-ENTHUSIASTS

Early adopters who embrace new technology immediately, eager to explore innovation. And probably not the person reading this book!

The rest of us in the queue hear about technology from these people first – its complexities, inconsistencies, and jargon.

The media speaks to Tech Enthusiasts first – often in jargon only they can follow. For everyone else, it's like overhearing a conversation in a language they don't speak, making new technology feel confusing rather than inviting.

But like everyone else, they are not without blind spots.

Characteristics

- Highly aware of digital trends
- Confident in navigating online spaces
- Trusts own ability to spot scams
- Enjoys testing new tools and platforms
- High risk appetite for online interactions

Potential Vulnerabilities

- Overconfidence – May assume they're immune to scams and fail to double-check sources
- Too trusting of innovation – More likely to fall for sophisticated scams mimicking new tech
- Less likely to question convenience – May overlook subtle warning signs

Outlook

Tech Enthusiasts thrive on innovation. They're the first to dive into new platforms, tools, and trends – often before the rest of us have even heard of them. But this eagerness comes with exposure to risks that haven't yet been fully understood or addressed. Technology itself isn't their weakness; being human is. Scams that exploit age-old vulnerabilities – like trust, urgency, or flattery – can bypass even the most advanced digital know-how. As we'll explore later in the book, enthusiasm without scepticism can be a dangerous mix.

TECHNOLOGY PERSONA 2: TECH-WILLING

Curious, thoughtful, and ready to engage, especially when they see clear value or hear positive experiences from people they trust.

The Tech-Willing are the first wave of mainstream adopters. They're not chasing the latest gadget, but they're not ignoring it either. They've seen what the early adopters are excited about and are receptive to trying new tools – once they've become affordable and available.

Characteristics

- Comfortable but selective with tech use.
- Follows mainstream digital habits.
- Open to change when it adds value.
- Can navigate digital options with guidance
- Trusts familiar platforms & recommendations

Potential Vulnerabilities

- Trust and Influence – Assumes scams are unlikely to appear on trusted platforms or from sources that seem credible, such as popular services or peer recommendations.
- Exposure to emerging threats – By engaging with new technologies early in their mainstream life, they're more likely to encounter scams that exploit novelty, limited awareness, or unrefined security.
- Hesitation to verify – May compromise security prompts in favour of convenience.

Outlook

Tech Willing users sit in the middle of the adoption curve. They're not first in line, but they're not far behind. Their openness is a strength – but only when paired with awareness. Trusting others to vet technology can work well, but it also creates blind spots.

Without a basic understanding of digital safety, they may unknowingly walk into traps disguised as convenience. Empowerment for this group comes from learning just enough to ask the right questions and spot the wrong answers.

TECHNOLOGY PERSONA 3: TECH CAUTIOUS

Only uses technology when necessary but prefers familiar systems and avoids unnecessary changes.

Products need to be mature before they commit. They're finally convinced by the people directly ahead of them in the queue – Tech Willing – but they usually wait a release or two later before adopting.

Characteristics

- Uses technology but with hesitation
- Prefers guidance & clear instructions
- Dislikes sudden or unexpected tech changes
- Feels overwhelmed by too many digital choices
- Avoids unnecessary online exposure

Potential Vulnerabilities

- Misplaced focus on technical skill – believing scam resilience requires deep tech knowledge, and may overlook the importance of habits, instincts, and simple checks.
- Easily pressured – Vulnerable to urgency-based scams due to hesitation in decision-making.
- Avoids verification steps – May assume unfamiliar warnings or messages are legitimate.

Outlook

To be cautious is to walk a fine line – balancing protection with empowerment. It's a thoughtful stance that helps you avoid unnecessary risks while still engaging with the digital world. But caution must be active, not passive. The danger lies in drifting too far from that line – disconnecting out of fear or uncertainty – and slipping into avoidance. When caution turns into shutdown, it becomes Tech-Terrified.

Staying informed and aware keeps you safely on the line, where you can protect yourself *and* enable yourself.

TECHNOLOGY PERSONA 4: TECH TERRIFIED

The resisters. Avoids technology as much as possible, feeling overwhelmed, vulnerable, or mistrusting of digital systems.

These people are last in line – by the time they adopt technology, it's no longer considered new and is well-established in everyday life. In fact, they're probably not in the queue at all. Some may never adopt certain innovations whatsoever.

Characteristics

- Avoids technology when possible
- Feels overwhelmed by digital interactions
- Fears making mistakes online
- Disengaged from cybersecurity awareness
- More likely to seek help for digital decisions

Potential Vulnerabilities

- Highly vulnerable to deception – Prone to following scam instructions out of confusion or fear.
- Relies on external guidance – May trust the wrong sources when seeking help.
- Easy target for authority scams – More likely to comply with scam messages from "official" sources.

Outlook

Their instinct to shut down or avoid technology altogether can feel protective – but it often leaves them uninformed and unprepared. This disengagement means they miss out on basic digital safety knowledge, making them vulnerable to scams that exploit fear, urgency, or authority. By staying disconnected, they lose the chance to build confidence and take control.

TECHNOLOGY CHANNELS

Phone Scams

CATEGORY	DETAILS
Scam examples	▪ Robocalls ▪ Fake bank or government calls ▪ Tech support scams
Red Flags	🔥 Applying pressure to act quickly 🔥 Applying fear or intimidation 🔥 Unknown Caller
Top 3 Protections Tips	1. Reject unknown numbers 2. Block and report suspected scams 3. Hang up if the caller applies fear or urgency pressures

Social Media Scams

CATEGORY	DETAILS
Scam examples	▪ Romance Scams ▪ Investments Scams ▪ Impersonation scams (Pretending to be someone you know)
Red Flags	🔥 Strangers claiming to know you 🔥 Requests for money, help or job referrals 🔥 Profiles with few posts or limited details
Top 3 Protections Tips	1. Limit what you share publicly 2. Verify connections before engaging 3. Report fake profiles or suspicious messages

Email Scams

CATEGORY	DETAILS
Scam examples	■ Phishing Emails ■ Invoice Scams ■ Job offer scams
Red Flags	🔥 Urgent payment requests or fake invoices 🔥 Attempts to apply intimidation or urgency 🔥 Generic Greetings (eg "Dear Customer")
Top 3 Protections Tips	1. Verify request through other channels 2. Hover over links to preview website links 3. Block and report suspicious emails

Fake Website Scams

CATEGORY	DETAILS
Scam examples	■ Fake Retail stores ■ Tech Support Pop-ups ■ Survey Scams
Red Flags	🔥 Search the web for the website or company name (add the word 'scam' to this to see if it gets any hits) 🔥 Pop-ups or links to download files 🔥 Unusual or misspelled website name
Top 3 Protections Tips	1. Verify requests through other channels 2. Hover over links to preview website links, ensure correct spelling 3. Block pop-up webpages in your browser settings

SMS Scams

CATEGORY	DETAILS
Scam examples	■ Delivery Scams (Auspost or Couriers) ■ Bank Alert Scams ■ Prize / Competition Scams
Red Flags	🔥 Requests for payment or to provide information 🔥 Attempts to apply intimidation or urgency 🔥 Sender listed as 'Unknown' or not a saved contact
Top 3 Protections Tips	1. Verify request through other channels 2. Don't click on links in SMS messages 3. Block and report suspicious emails

Fake Phone Apps Scams

CATEGORY	DETAILS
Scam examples	■ Banking App clones ■ Flashlight/Utilities that request to much permissions ■ Crypto Wallet Scams
Red Flags	🔥 Attempts for you to download apps from a link in email, SMS or on a website 🔥 Apps that seek excessive permissions 🔥 Apps with bad reviews or no reviews
Top 3 Protections Tips	1. Install and update apps from your App Store only (Google Play Store or Apple App Store) 2. Be wary of apps with few reviews or bad reviews 3. Block and report suspicious emails

REFERENCES

Introduction

[1] Global Anti-Scam Alliance (GASA) (2024, November 7) International Scammers Steal Over $1 Trillion in 12 Months in Global State of Scams Report 2024. https://www.gasa.org/post/global-state-of-scams-report-2024-1-trillion-stolen-in-12-months-gasa-feedzai

[2] Australian Human Rights Commission. (2025, August 19) Statistics about technology and human rights. https://humanrights.gov.au/our-work/education/stats-facts/statistics-about-technology-and-human-rights

[3] Australian Cyber Security Centre. (2023, April 21) What are Scams? Cyber.gov.au https://www.cyber.gov.au/threats/types-threats/scams

[4] Australian Cyber Security Centre. Cyberattack. Cyber.gov.au. https://www.cyber.gov.au/glossary/cyberattack

Chapter 1

[3] Scamwatch. Scam Statistics. Australian Competition and Consumer Commission. https://www.scamwatch.gov.au/research-and-resources/scam-statistics

Chapter 3

[1] IDCARE (2024). 'Hı Mum' scam continues to trick parents three years on from first texts reported in Australia.
Retrieved from: https://www.idcare.org/news/the-national-scam-identity-and-cyber-support-servi

Chapter 4

[1] The Associated Press / FRONTLINE (2025, December 17). Myanmar Declares a 'Zero Tolerance' Policy for Cyberscams. But the Fraud Goes On. https://www.pbs.org/wgbh/frontline/article/myanmar-cyberscam-scam-compound/

[2] Centre for Information Resilience – Myanmar Witness (2025, November 25). Reported Explosions, Shutdowns and the Raid at

KK Park. https://www.info-res.org/myanmar-witness/articles/
reported-explosions-shutdowns-and-the-raid-at-kk-park/

[3] The Irrawaddy (2025, December 8). Scam Centres Flooding Myawaddy
Town After Myanmar Junta Crackdown. https://www.irrawaddy.com/
news/burma/scam-centres-flooding-myawaddy-town-after-myanmar-
junta-crackdown.html

[4] UN OHCHR (2023, August 29). Hundreds of Thousands Trafficked to
Online Scam Centres in Southeast Asia, says UN Report. https://www.
business-humanrights.org/en/latest-news/southeast-asia-ohchrs-
report-reveals-hundreds-of-thousands-of-victims-held-in-online-scam-
operations-in-southeast-asia-with-cambodia-myanmar-as-main-hubs/

[5] INTERPOL (2025, June 30). New Information on the Globalization of
Scam Centres. https://www.interpol.int/en/News-and-Events/News/2025/
INTERPOL-releases-new-information-on-globalization-of-scam-centres

[6] INTERPOL (2024, June 27). USD 257 Million Seized in Global Police
Crackdown Against Online Scams — Operation First Light 2024. https://
cybernews.com/news/interpol-global-crackdown-scams/

[7] Stamp Out Scams (2025, August 10). Inside the Shadows: Scam Farms,
Human Trafficking, and the Unseen Victims of Southeast Asia's Digital
Crime Wave. https://stampoutscams.org/inside-the-shadows-scam-
farms-human-trafficking-and-the-unseen-victims-of-southeast-asias-
digital-crime-wave/

[8] National AntiScam Centre (2025, March 11). Australians Better
Protected as Reported Scam Losses Fell by Almost 26 Per Cent.
Scamwatch / ACCC. https://www.scamwatch.gov.au/about-us/news-
and-alerts/australians-better-protected-as-reported-scam-losses-fell-by-
almost-26-per-cent

Chapter 5

[1] Byrne, D. (1971). The attraction paradigm. Academic Press. Berscheid,
E., & Walster, E. (1969). Interpersonal attraction. Addison-Wesley.

[2] Global AntiScam Alliance (GASA). (2025, October 7). Global State of
Scams 2025 Report. https://www.gasa.org/post/global-scams-on-the-rise-
over-half-of-adults-worldwide-report-scam-encounters

[3] National AntiScam Centre (2025, March 11). Australians Better
Protected as Reported Scam Losses Fell by Almost 26 Per Cent.
Scamwatch / ACCC. https://www.scamwatch.gov.au/about-us/news-
and-alerts/australians-better-protected-as-reported-scam-losses-fell-by-
almost-26-per-cent

ACKNOWLEDGEMENTS

This handbook started in a community workshop when someone asked, *"Is there a handout?"* That question sparked the idea to create something practical for building confidence online.

True to the theme of the book – privacy matters, even in acknowledgements. So, no names – just gratitude to the following:

Thank you to everyone who attended my workshops and shared the message of online safety. You kept this project moving forward.

Community defence is crucial and I have been blessed with an incredible one.

To my ICT community in Tasmania and colleagues across Ireland, the UK, and Australia – thank you for three decades of support and for tolerating my analogies about technology being like houses and cars. Those comparisons helped me explain complex ideas in simple ways, even if they rolled a few eyes over the years.

To friends, neighbours, and workshop attendees who proofread drafts – your sharp eyes made this better.

Finally, to my closest community: my family. The one I was born to and the one I built. To my wife and our two sons. Please remember the safe word…

'RED FLAGS CHECKLIST: QUICK ACCESS

RED FLAG	WHAT TO WATCH FOR	CHECK?
Unknown Contact	Unknown number, unsaved contact, unfamiliar person, odd email address	☐ Yes ☐ No
Urgency	"Act now", countdowns, pressure to act before verifying	☐ Yes ☐ No
Fear	Threats of legal action, account suspension, or fines	☐ Yes ☐ No
Context	Requests via SMS/email instead of official app or website	☐ Yes ☐ No
Payment	"Pay now", gift cards, crypto, wire transfers	☐ Yes ☐ No
Links	Login pages, short web links, 'Click here' buttons	☐ Yes ☐ No
Confusion	Baffling you with technical jargon, financial complexity or something you don't understand	☐ Yes ☐ No
Offers	Promises of free money, prizes, exclusive deals	☐ Yes ☐ No
Emotion	Guilt trips, sympathy ploys, help requests	☐ Yes ☐ No
Errors	Poor grammar, typos, incorrect logos, wrong provider name, not a service you use	☐ Yes ☐ No
Unexpected	A charge, email correspondence or interaction you are not expecting	☐ Yes ☐ No